SCIENCE AND RUSSIAN
CULTURE IN AN AGE
OF REVOLUTIONS

SCIENCE AND RUSSIAN CULTURE IN AN AGE OF REVOLUTIONS

V. I. Vernadsky and His Scientific School, 1863–1945

KENDALL E. BAILES

Indiana University Press
Bloomington and Indianapolis

This book was brought to publication with the assistance of a grant from the Andrew W. Mellon Foundation to the Russian and East European Institute, Indiana University, and the Center for Russian and East European Studies, University of Michigan.

Manufactured in the United States of America

Library of Congress Cataloging-in-Publication Data

Bailes, Kendall E.
Science and Russian culture in an age of revolutions :
V.I. Vernadsky and his scientific school, 1863–1945 / Kendall E.
Bailes.
p. cm.—(Indiana-Michigan series in Russian and East
European studies)
Bibliography: p.
Includes index.
ISBN 0-253-31123-3
1. Vernadskiĭ, V. I. (Vladimir Ivanovich), 1863–1945.
2. Geologists—Ukraine—Biography. 3. Science and state—Soviet
Union. I. Title. II. Series.
QE22.V47B34 1990

509.47—dc20

89-45358
CIP

1 2 3 4 5   94 93 92 91 90

CONTENTS

Foreword

This book is the product of two dramatic stories, both of which contain elements of triumph and tragedy. The first story concerns the subject of the book, Vladimir Ivanovich Vernadsky. The second story is about its author, Kendall E. Bailes.

Vladimir Vernadsky was at the time of the Russian revolution already fifty-four years old, a distinguished geologist and a member of the Imperial Academy of Sciences, a man whose personality and political views were fully formed. Yet he lived on as an active scientist under the new Soviet government for almost another entire generation, dying at the end of World War II. As a liberal in politics and an eclectic in philosophy, Vernadsky never felt comfortable with either of the two systems of government that occupied almost all of his life, the tsarist empire that fell in early 1917 and the Soviet government that arose toward the end of that year. Under both systems Vernadsky feared for the fate of science and learning, and under both he was threatened with dismissal and arrest.

Nonetheless, Vernadsky built a distinguished career before the revolution and expanded on it after that event. He was not devoid of ambition, but he combined it with a commitment to serving science and the people of his country. Fluent in several languages and a frequent traveler abroad, he could easily have emigrated, following the example of his son George, later to be a distinguished historian of Russia at Yale University. He stayed on, however, and, through his knowledge of geology, helped the Soviet government's industrialization effort. Vernadsky wanted Russia to be strong even if its government was distasteful to him; that strength, after all, would be inherited later by succeeding forms of government. Vernadsky's liberal faith led him to believe that a more democratic government lay somewhere in the future.

A number of scholars familiar with the separate destinies of the father and the son have assumed that they differed politically. Some differences about politics and, especially, religion did exist between the two, but it would be a mistake to exaggerate them. The different disciplines of the geologist Vladimir and the historian George already implied different political choices. Both Vladimir and George knew that a liberal historian could not survive in Stalin's Russia, while a liberal geologist might.

Vladimir Vernadsky was not content, however, merely to confine himself to geology, thus fleeing political difficulties. As Kendall Bailes shows in this book, he strove to develop an alternative philosophical viewpoint to the dogmatic Marxism that Soviet ideologists were propagating. Although Vernadsky was not a militant fighter for causes, he stoutly protested political incursions on academic freedom. He was, for example, the main organizer of resistance to the Communist takeover of the Academy of Sciences in the late 1920s, delivering speeches and writing memoranda about the damage that politics could do to science. He lost that battle in the sense that the takeover succeeded, but he won by keeping alive the spirit of independence of the full members of the Academy, especially among the natural scientists who were less vulnerable than the humanists and social scientists. That spirit survived and has emerged again and again, most dramatically in the person of Andrei Sakharov, a member of the Academy whose views on political liberalism and the importance of international cooperation in science fully accord with those of Vernadsky.

During Vernadsky's life he was regarded by the Soviet authorities as a valuable geologist with untrustworthy political views. Twenty years after his death a much more positive picture of Vernadsky emerged in Soviet publications. Soviet authors praised his contributions to geology, his prescient concepts of the biosphere and noosphere, and, after the advent of Gorbachev to power, even his pleas for political tolerance. At present Vernadsky is a great figure in Soviet intellectual life, as attested by a constant stream of publications by and about him.

In none of these publications, however, are the richness and heterogeneity of Vernadsky's life or the ambiguity of the issues he faced portrayed as fully as in Kendall Bailes's book. In the first three chapters, in particular, Bailes displays a rare sympathetic understanding of Vernadsky's boyhood and student years and his struggles as a young professor. The book is destined to become a valuable contribution to our knowledge of the history of Russian science and to our understanding of Russian and Soviet culture.

In the first paragraph of this foreword, I commented that this book embodies a story of triumph and tragedy not only about Vernadsky but also about Bailes. I am writing this foreword because Kendall died of AIDS before he could write it himself. Sad as that necessity is, it gives me an opportunity to express the appreciation of his colleagues for his remarkable contribution to Soviet history and the history of science and technology.

In a short professional life of less than two decades, Kendall produced three books and about twenty articles. His book *Technology and Society under Lenin and Stalin: Origins of the Soviet Technical Intelligentsia* (Princeton, 1978) has become a classic work in Soviet history, one of only a handful of books in the field that deserve that description. Winner of the Herbert Baxter Adams Prize of the American Historical Association in 1979, it has earned acclaim not only from historians of the Soviet Union but also from historians of science and technology working on topics in Western Europe and North America. In this work Kendall produced a synthesis of the history of technology and political history. His interpretation of the role of technocracy in the 1920s adds a major insight to our understanding of the Soviet Union at the beginning of the Stalinist period.

Kendall's scholarly interests transcended Soviet studies, as shown by the book he edited on environmental history (*Environmental History: Critical Issues in Comparative Perspective*, American Society for Environmental History, University Press of America, 1985). Long on the faculty at the University of California in Irvine, he was involved in efforts there to preserve the environment.

When Kendall learned that he was suffering from AIDS, several of his friends and colleagues asked him if there was any way they could help him complete his book on Vernadsky, which had already occupied him for a number of years. He replied that he wanted to finish it himself if he could, but that he would appreciate knowing that if he did not succeed, his friends would step in to complete the task. He also said that he needed financial and research assistance. Five historians specializing in the history of Russian and Soviet science and technology—Mark Adams, Harley Balzer, Loren Graham, Paul Josephson, and Douglas Weiner—formed an unofficial committee for this purpose. Janet Rabinowitch of Indiana University Press provided encouragement and helpful advice about publication. The MacArthur Foundation, the Ford Foundation, and the Joint Committee on Soviet Studies of the Social Sciences Research Council came to Kendall's assistance with research grants. Lori Citti, a talented and generous graduate student from Indiana University, became Kendall's research assistant. She worked closely with Kendall during the last year of his life. Her magnanimity inspired us all.

I visited Kendall in Los Angeles a few weeks before his death. He was confined to a bed in his apartment just off Hollywood Boulevard and was being cared for by his parents, Vivian and Ira Bailes, and Lori Citti. I did not know how close to completion the manuscript was and feared to ask. Kendall was mentally alert but more interested in talking of his worries about the degradation of the natural environment of California than of his own declining health or his book on Vernadsky. He had recently written several poems about his environmental concerns that he read to us. I left without knowing exactly what the state of the book was, although Lori Citti assured me that it was progressing well.

After Kendall's death Lori mailed the manuscript to Janet Rabinowitch and the unofficial committee. To our delight, we found that it was entirely complete and that it was a work of scholarly importance. Ken met his last deadline. We will continue to benefit from his sense of duty and from his lively mind. We miss him as a friend and colleague.

Loren Graham

Postscript. I would like to express special thanks to Douglas Weiner, who during the publication process took time from his own work to check the edited manuscript and proofread the galleys.

Introduction

After 1956, when Nikita Khrushchev made his secret speech about Stalin and controls over intellectual life loosened somewhat, small groups of Soviet intellectuals, centered especially in the Academy of Sciences, began to rediscover and promote Russian thinkers who for political reasons had been neglected during the later Stalin years. Two of these neglected thinkers whose work was revived and popularized in the 1960s and 1970s were Mikhail Bakhtin (1895–1975) and Vladimir Vernadsky (1863–1945). In recent years, Vernadsky has become something of a popular figure in the USSR. Many of his works have been published, popular biographies and television programs about him have been created, and his face peers out at contemporary Soviet citizens from numerous public portraits, statues, book covers, postage stamps, and memorial lapel pins. One of the major new boulevards and a Metro station in Moscow bear his name. In other words, he has become something of a Soviet cultural icon, although the precise meaning of this icon for the ordinary Soviet citizens and for Soviet intellectuals may differ. For ordinary citizens, Vernadsky's popular image has come to symbolize a great Russian patriot, one of the major Soviet scientists of the twentieth century. For intellectuals, he has come to have a different and more specialized meaning, closely related to the frustrations felt by Soviet scientists and other scholars at the limitations placed on their professions by the Soviet state and the Communist party.

At first glance one would think that a literary critic, theorist, and philosopher of language such as Bakhtin would have little in common with an earth scientist, geochemist, and philosopher of natural science such as Vernadsky. Yet the two share an important affinity in their view of knowledge, and I would argue that it is this affinity that accounts in part for their espousal by influential circles of Soviet intellectuals, who felt frustrated by the scholasticism and dogmatic certainty of official philosophy under Stalin. These Soviet intellectuals sought to change the official philosophy of the Soviet Union, dialectical materialism, by absorbing into it a different view of knowledge and how knowledge is obtained, tested, and viewed in Soviet intellectual life. I argue in this book that such Soviet scholars have been working over the past two decades or so to strengthen in Russian culture an ideology of professionalism, one with strong native Russian roots, which will help to protect their freedom of inquiry, i.e., their freedom to debate and disseminate ideas without arbitrary interference by political authorities. I will attempt to show, through an analysis of the career and work of Vladimir Vernadsky and his scientific school, the social matrix in which such ideas originated and why they have become popular among a later generation of Soviet natural scientists and other intellectuals. The affinity with Gorbachev's group and *glasnost* should be obvious. I am told that Gorbachev quoted Vernadsky in one of his speeches, but I have been unable to find the reference. Certainly the intellectuals in the Academy of Sciences who are Gorbachev's allies, such as Evgenii Velikhov and Academician Roald Sagdeev, are familiar with Vernadsky, one of the "heroes" of Russian science. Even a recent Soviet novel by Daniil Granin, *Zubr*, portrays Vernadsky as a figure of hero worhip for younger scientists and intellectuals.

But let me return to the comparison between Vernadsky and Bakhtin. Although Bakhtin's ideas about knowledge are difficult to sum up, he maintained that no system of knowledge is final or absolutely true; there are no ultimate explanations that everyone, without exception, will accept as exhausting all possibilities.

Bakhtin, of course, is not alone in recognizing the heterogeneity and contradictions that dominate human thought and the speciousness of all claims to absolute knowledge. Unlike many others, he does not find the absence of certainty about our knowledge a matter to deplore. His uniqueness resides in the manifold ways he celebrates and specifies the world's unpredictability and the necessity of constant free dialogue among people involved in trying to create meaning out of the messiness and uncertainty of a world in flux. His suppressed books and unpublished manuscripts, written in the 1920s and 1930s for the most part, began to be published in the Soviet Union in the 1960s and 1970s. They formed a strong argument for the freedom of all people to engage in dialogue and create their own meanings from such dialogue. His writings also formed a strong argument against the belief that absolute truth is possible, since Bakhtin believed that the search for such certainty reinforces oppressive political and religious systems. His philosophy of dialogism was an argument against monologic belief systems, which hold that a single truth exists and can be contained and guarded in institutions such as a state, a political party, or an official church. In the past five years, Bakhtin's writings have become accessible to Western scholars through the translation of his dissertation on Rabelais and a recent biography by Michael Holquist and Katerina Clark.

The eminent natural scientist Vladimir Vernadsky held a similar view of knowledge, and during the 1930s he wrote several substantial books about the history and philosophy of the natural sciences in which he attempted to show that all knowlege about nature should be viewed as partial, tentative, and incomplete. Scientific knowledge, in Vernadsky's view, could only grow and change by allowing schools of scientists the freedom to argue, debate, and constantly subject their theories to empirical tests and the judgment of a heterogeneous scientific community, unhampered by the interference of church, state, or political party. Like much of Bakhtin's work written during the Stalin era, Vernadsky's philosophical works were not published until the 1960s and 1970s.

In both cases, the publication of long-suppressed manuscripts written in the twenties, thirties, and early forties was sponsored by influential groups of scholars in the Soviet Academy of Sciences, with the help and patronage of important Communist party officials and ideologists.

This seems strange at first glance, for neither Bakhtin nor Vernadsky was a Communist party member or sympathizer. Both, in fact, had some difficulties with the regime during the Stalin years, Bakhtin more so than Vernadsky. Bakhtin, for example, was arrested and exiled from Leningrad in 1930 as a "class alien element." (His father was a nobleman and bank official before the revolution, and Bakhtin, although strongly anticlerical, was affiliated with Russian Orthodox intellectual circles in Leningrad during the 1920s.) However, it should be noted, he also drew ideas from Marxism and the natural sciences and considered himself a religious and philosophical materialist at the same time, following the views of such religiously oriented Russian natural scientists as Ivan Pavlov and A. A. Ukhtomsky, another biologist with whom Bakhtin was closely associated in the 1920s. After his release from internal exile in the later 1930s, Bakhtin was allowed to teach Russian literature at a teachers' college in Soviet Mordovia, a bleak area in northern Russian full of former political exiles. He was also allowed to finish his doctoral dissertation at the Gorky Institute of World Literature in Moscow, but he was not allowed to publish any of his voluminous manuscripts written in this period. Then in the late 1950s, during the Khrushchev years, a group of young literary scholars at the Gorky Institute rediscovered Bakhtin's book from 1929 on Dostoyevsky, published just before his arrest, and his unpublished doctoral dissertation on Rabelais in the archives of the Gorky Institute. Believing at first that Bakhtin was dead, they attempted to have these works published; and then, discovering that he was still alive, having just retired as a professor at this obscure teachers' college in Mordovia, they sought him out, befriended him, and began to wage a long and difficult campaign to have his works published in the Soviet Union. They were helped in this campaign by such establishment writers and Communist party literary functionaries as B. Riurikov and Konstantin Fedin. Even the daughter of Iurii Andropov intervened with her father, the head of the KGB at the time, to help Bakhtin. She was a graduate student in the Gorky Institute and a follower of one of the professors who had helped rediscover Bakhtin. This campaign to publish the works of Bakhtin was opposed in the press by conservative Stalinist literary theoreticians such as Aleksei Dymshits, but the tide went against them. Starting in 1965, the proponents of Bakhtin's brilliance succeeded in their campaign to rehabilitate him; and one after another in the late 1960s until his death in 1975, a series of major books and essays by Bakhtin became a literary sensation, first in Moscow, then in Paris, and in recent years among literary critics and theorists in the United States.

In 1981, while I was working as a visiting scholar in the Institute for the History of Science and Technology in Moscow, I discovered that a similar process had taken place in the 1960s and 1970s in rehabilitating and popularizing the unpublished works of Vladimir Vernadsky. I had been interested in the career of Vernadsky and his scientific school since the early 1970s, when I first met his son George Vernadsky, a historian at Yale. I was struck by the flood of books by and about his father, both scholarly and popular, being published in the Soviet Union at the time. What I learned about this in Moscow in 1981 was that as early as 1957 a circle of some twenty scientists had been formed in the Vernadsky Institute of Geochemistry located in the Lenin Hills near Moscow University. They began to meet regularly to study his intellectual legacy, including his unpublished manuscripts located in the Academy archives. Although at first opposed by the director of the Institute, this group enjoyed the high-level protection of Iurii Zhdanov, an organic chemist and important administrator of science who was the son of Andrei Zhdanov, party boss of Leningrad in the later 1930s and a politburo member until his death in 1948. One of Iurii Zhdanov's close allies in this effort to rehabilitate Vernadsky was Bonifatti M. Kedrov, an important philosopher of science who at the time was director of the Institute for the Study of Science and Technology in Moscow. Like Zhdanov, Kedrov had high party connections. He was the son of an Old Bolshevik and medical doctor. Kedrov's father, in fact, was one of the early leaders of the Bolshevik party before 1917 and one of the founders of the Soviet state, including the political police in which he served during the Russian civil war under Dzherzhinsky. Kedrov's father, like so many Old Bolsheviks, was executed during the Stalinist purges of the later 1930s and then rehabilitated posthumously after 1956. Kedrov became a major opponent of Lysenkoism as a party philosopher of science in the later 1940s and 1950s and has been known for many years as a party liberal. Iurii Zhdanov, who had been appointed by his father as head of the Department of Science for the Central Committee of the Communist party for a time after World War II, when he was also married to Stalin's daughter, later became rector of an important university and an influential official in the Academy of Sciences. He and Kedrov signed off as responsible editors on most of Vernadsky's suppressed works published in the 1960s and 1970s. Closely associated with them were a group of earth scientists, historians, and philosophers of science in the Academy, including the man who is now head of the Geology-Earth Sciences Section of the Academy, Aleksandr Leonidovich Ianshin. Through people associated with this group, I obtained access to many unpublished documents from the Vernadsky archives. They also helped considerably in understanding the social conditions of Soviet science in which Vernadsky worked until his death in 1945 and the conditions that led to a revival of interest in his career and ideas after Stalin's death.

SCIENCE AND RUSSIAN CULTURE IN AN AGE OF REVOLUTIONS

V. I. Vernadsky, 1940.

CHAPTER
1

A LIFE FOR SCIENCE
Vernadsky's Development as a Scientist in Imperial Russia, 1863–1888

"Thinking about my life," Vladimir Vernadsky wrote to his son in 1944, the year before his death at the age of eighty-three, "I see a very close connection between generations. In our family the past life of our predecessors can be felt down to the present time."[1] In the years prior to his death, Vernadsky had been thinking a good deal about family history and was gathering material for his memoirs, which he was never able to complete. But thanks to the materials which he and his son collected, we are able to understand what he meant by the "very close connection between generations." Vernadsky's family background contributes to our understanding of his later thought and career as a scientist and public figure.

The biographies of many members of the Russian intelligentsia in the nineteenth century were marked by rebellion against parental authority and parental values and often against the values of older intellectuals as well. When such individuals came from a family of the intelligentsia, this could involve a dual rebellion against their elders both as parents and as representatives of an older generation of *intelligenty*. Vernadsky was born in 1863 during the controversy over Turgenev's novel *Fathers and Sons*, which personified one writer's view of the conflict between generations of Russian intellectuals, in this case between an older generation of liberals, who valued the arts and humanities as well as natural science and were often better at talking about reforms than acting, and a younger generation of radicals, whom Turgenev dubbed nihilists. Nihilists, or "realists" as they preferred to call themselves, believed in subordinating the arts and humanities to the natural sciences and to utilitarian purposes. They derided the ineffectuality of older intellectuals in bringing about social change. Their views were closely related to French positivism, and they often chose careers as natural scientists or as revolutionaries, sometimes as both, in defiance of their elders among the intelligentsia and as part of their aim of overturning the entire structure of "official" Russia, most particularly the Tsarist bureaucracy and Old Regime society.[2] However, this was not to be the pattern of Vernadsky's life. He came to the natural sciences by a different path, and he was an active opponent of the nihilist (and positivist) attitude toward the natural sciences and toward culture in general. In his attitude toward social change, he was an evolutionist, a believer in gradual change who placed his faith in the transforming power of modern rationalist culture rather than in a sudden and revolutionary change in the social structure.

Vernadsky became a rebel against Tsarist authority, but a rebel from within. Although he chose a career as a natural scientist, he believed not only in the natural sciences *(estestvoznanie)* but in *nauka,* the pursuit of organized knowledge in the widest sense, similar to the German concept of *Wissenschaft*. From his youth he read widely in philosophy, history, and literature and wrote extensively on the first two topics in later life. Like his father, he became one of Russia's foremost liberal intellectuals. In fact, he never seems to have rebelled against the values of his father, who represented an older generation of Russian intellectuals. Quite the contrary, a strong sense of continuity in values and beliefs runs across the generations. Although it is always difficult to untangle the scattered threads that influence the formation of an individual, there are a number of interests, beliefs, and patterns of behavior that Vernadsky first encountered in the family setting. The family and, then later, Tsarist schools served as the nexus where broader currents in Russian life met in the two decades between his birth and graduation from St. Petersburg University (1863–85). Vernadsky's family history and his schooling will illuminate the origins of his values and thought. Such discussion may also help to explain why Vernadsky was able to bridge successfully two such seemingly distinct eras as those of Tsarist Russia and the Soviet period without changing his basic values. The patterns of interaction with official authority which he learned as a youth, both in his family and at school, are crucial to an understanding of his later behavior in Imperial Russia and under the Soviet regime.

Vernadsky's family came from the hereditary service class of Imperial Russia, where they held the status of hereditary nobles. The term "service class" is used interchangeably with that of hereditary nobility, although the latter term can be somewhat misleading. For most Westerners, "hereditary nobility" has connotations often inappropriate to the Russian situation; it conjures up images of landed estates passed down through a family for generations, a coat of arms perhaps, and a well-verified pedigree, often rooted in a single locality for hundreds of years. The Russian service class, by contrast, was generally more fluid, often landless, more insecure and dependent on the centralized state, of which it was in fact the product, and less rooted in a single locality for any lengthy period. Particularly in the middling and lower ranks, members of the service nobility had little or no landed property and lived from the income of government service. It was preeminently a bureaucratic class, created to serve the needs of the state, and continually renewed from the ranks of non-nobles.

The ethnic and religious origins of the service class were therefore hetero-
genous, reflecting the ethnic diversity of the Russian Empire itself and the large
number of foreigners brought in over the centuries to serve the tsars.[3]

The family into which Vladimir Vernadsky was born in 1863 was typical of the
Russian service class in these respects. To be sure, Vernadsky's father had
acquired some land, although the family rarely resided there; but he was
financially dependent on non-agricultural pursuits for most of his life. On the
Vernadsky side, the origins of the family tree are not well established before the
eighteenth century. The Vernadsky family had apparently been "free Cossacks,"
members of the *Zaporozhe Sech* in the Ukraine.[4] The Cossacks preserved a
strong family tradition of a more democratic social organization before they
were subdued and their way of life changed by the tsarist government in the
eighteenth century. The family name, originally spelled Vernatsky, may have
been Lithuanian, but its origins are not known for certain.

At any rate, Vernadsky's great grandfather, Ivan Nikiforich Vernatsky, left the
Cossacks and went to a theological seminary in the Ukraine. Married to the
daughter of another Cossack family whose origins were Polish, he later settled
down in a village in the Ukraine, where he was elected parish priest by a local
Orthodox congregation. The election of priests by local parishioners and parish
autonomy in important matters were still common in this region and had not yet
been replaced by stronger control from above by the Orthodox Church hier-
archy. In 1813, Ivan Nikiforich also had his name entered into the ranks of the
local nobility, on the sworn testimony of twelve local noblemen that he came
from the ranks of the "free Cossacks," lived "in the manner of a nobleman," and
therefore had a legitimate claim to noble status.[5] By permitting "free Cossacks"
to attain nobility, the government hoped to quell rebellious tendencies in its
frontier regions and turn the Cossacks into staunch defenders of the autocracy.
Since the Orthodox clergy formed a separate and subordinate legal category in
the Russian Empire,[6] there were obvious advantages to being a nobleman. For
example, the nobility was exempt from taxation and corporal punishment. How
Ivan Nikiforich managed to remain both a nobleman and a priest is baffling and
only illustrates the mixed nature of Russian social groups in this period, the
overlapping of legal categories, and the lack of a firm rule of law in Tsarist
Russia. The marginal nature of noble status for many families is reflected in the
fact that the Vernatsky family's claim to nobility was questioned in the 1840s,
when Nicholas I attempted to cull from the ranks of this estate persons of
dubious status and property holdings. Although in this period most of his
relatives in Chernigov Province, the Ukraine, were stricken from the ranks of
nobility, the son of Ivan Nikiforich, Vasilii, had by then earned hereditary
nobility through his service as an army doctor, that is, by achieving a service
rank that conferred noble status on himself and his descendants.[7]

With Vasilii, the grandfather of the scientist, more secular and scientific
interests entered the family. Vasilii was a rebel against his father's values in a
way his grandson was not. His father had wanted him to enter a church

seminary and follow a career as a priest, but Vasilii preferred to enter the University of Moscow. He went to Moscow on foot, according to family legend, with his mother's secret blessing, and although he was unable to enter the university there, he did enroll in the Medical-Surgical Academy. His defiance of paternal authority earned him his father's notorious anger. Ivan Nikiforich placed a curse on Vasilii and all his progeny.[8] As a result, Vasilii changed the spelling of the family name to Vernadsky and made the break with traditional village life more complete. Despite his service in the Imperial Russian Army, a strong element of sympathy for the Ukraine and for the Ukrainian language and culture remained a family tradition that later had considerable influence on his grandson, Vladimir. In general, as Vladimir Vernadsky later expressed it, writing to his fiancee in 1886, "Family traditions had a rather strong influence on me in childhood."[9]

Vasilii married into another Ukrainian service family, the Korolenkos. During the reign of Alexander I (1800–25), his father-in-law was the collector of customs at Taganrog, a port on the Sea of Azov near the Black Sea.[10] These same Korolenkos later produced the well-known writer and radical journalist of the late nineteenth and early twentieth centuries, V. G. Korolenko. In fact, the writer Korolenko and the scientist Vernadsky shared the same great grand-father, the Taganrog customs collector. V. G. Korolenko's popular memoirs about his family and life in late Imperial Russia, *The History of My Contempo-rary* (still a favorite in Soviet schools), were later to remind his second cousin, Vladimir Vernadsky, of many scenes from his own childhood.[11]

Vernadsky's grandparents had a large family, all of whom except one, how-ever, died as children. Believing that his father's curse was to blame for their misfortune, Vasilii and his wife, Ekaterina, decided to name their youngest son Ivan, after his grandfather, hoping to placate the wrathful spirit of the now deceased parish priest. Although Vasilii was a physician and generally a secular man, he still carried within him a superstitious streak that linked him to the traditional culture of his ancestors. Whatever the reason, Ivan not only lived to adulthood but also became famous as one of Russia's foremost liberal political economists in the mid–nineteenth century. It is with Ivan Vernadsky that we find the first strong evidence of active dissidence toward the Tsarist autocracy. The attitudes that led to such dissidence may well have originated in the army doctor's family. After the break with his clergyman father, the army doctor had become a Mason, evidence of free thinking but not necessarily of active dissi-dence. Vasilii was also sympathetic toward the Decembrists, those army officers and noblemen who tried to overthrow the autocracy in 1825,[12] but nothing in the historical record indicates that Vasilii was more than a closet opponent of autocracy. Whatever his attitude toward the autocracy, the army doctor had a strong reputation as a kindly and humanitarian person, qualities passed on to his son. To give one example, while serving as head of a field hospital with Suvorov's army during the Napoleonic Wars, Vasilii was captured by the French in Switzerland in 1799. Napoleon later awarded him the Legion of Honor for outstanding service to the wounded of both sides, before and during his captivity.[13]

Vasilii's son, Ivan, became known not only as a broadly tolerant and humanitarian person but as an active proponent of a constitutional and liberal regime in Russia, one where the despotic arbitrariness of the autocracy would be replaced by the rule of law and by a parliamentary system. His reputation for integrity was such that in Korolenko's book *The History of My Contemporary*, one of the characters from the period of the 1870s proclaims, "I know of only one decent person, yet only one in all of Russia—Ivan Vasilevich Vernadsky." Yet the same character goes on to complain that even Vernadsky tended to defer too much to Tsarist authority.[14] Russian liberals like Ivan Vernadsky often found themselves walking a tightrope between their desire for significant social and political changes and the need to give in at times to a state which, in the final analysis, still controlled their means of livelihood and their status in society.

Ivan Vernadsky was born in Kiev, the Ukraine, and finished the university there in the early 1840s. His father had died in 1836, when Ivan was only fifteen, leaving no property to speak of. He and his mother were helped by an uncle, M. Y. Korolenko, who had acquired some property in the Tsarist service. Ivan Vernadsky's original interests had been in Slavic languages and philosophy; his undergraduate thesis was on Platonic philosophy and the concept of the soul. Perhaps because of his family background, he was particularly interested in Polish and Ukrainian and was known to have special sympathies for these peoples. By the age of twenty-two, in 1843, Ivan Vernadsky had already spent several years teaching Russian at a secondary school. Because of the gold medal he won for his work as an undergraduate, he was chosen by his university in 1843 to pursue graduate studies in Western Europe. This was in preparation for a career as a university professor.[15]

Vernadsky applied for a vacancy in Slavic linguistics, but someone else was chosen for that position. The Ministry of Education offered him another vacancy, in the field of political economy, which at that time was just developing in Russia. Vernadsky was assigned the task of preparing himself in England and Germany to teach political economy and statistics, fields in which he eventually distinguished himself by authoring several books, including one of Russia's first textbooks on economic statistics.[16] After returning from Western Europe, Vernadsky completed two advanced degrees at Kiev University and was appointed to the chair of political economy. By the 1850s he had transferred to Moscow University.

About the same time he married Maria Shigaeva, one of Russia's early feminists and the country's first woman economist. In the spirit of the advanced intellectuals of the time, they settled down to create an equal and companionate relationship. By all accounts the marriage was successful, although it was cut short by Maria's early death, of tuberculosis, in 1860.[17] On her recommendation and with her help, Ivan founded a new journal, the *Economic Index*, which played an important role in the debates of the reform era, particularly on questions concerning peasant emancipation and women's rights.[18] Maria Vernadskaia wrote a number of popular articles for this journal, discussing the situation of women, advocating equal treatment and employment outside the home to make women less dependent economically on fathers and husbands. In his book *The Women's Liberation Movement in Russia*, Richard Stites called the Vernadsky journal the main advocate of Manchester liberalism in Russia at the time.[19] In a sense the *Economic Index* was just that, calling for free trade, free enterprise, and free labor. The Vernadskys promoted the emancipation of serfs without land (and therefore without heavy redemption payments), seeing in such an emancipation the creation of a large new force of free labor that would spur industrialization. This brought Vernadsky into conflict with liberals among the nobility who wanted the peasants freed, but required to buy a share of the land at high prices in compensation to owners for the loss of serf labor. Vernadsky saw high redemption payments, quite correctly as it turned out, as a major hindrance to industrialization and the modernization of Russian agriculture. After beginning publication in 1857, when the Vernadskys moved to St. Petersburg to be at the center of the publishing industry, the *Economic Index* also found itself involved in polemics on the question of the peasant commune with N. G. Chernyshevsky, the most important radical journalist living in Russia at the time and editor of *The Contemporary*. Vernadsky saw the rural commune *(mir)*, with its common landownership and collective responsibility, as a retrograde institution that served as a roadblock to economic development, whereas Chernyshevsky saw in its preservation the seeds of peasant socialism.

In most respects, Ivan Vernadsky could be called a Manchester liberal; but it is also clear from his journal that he never believed in the unbridled rights of private property or in a highly restricted state. In his view, as in that of his son in later years, the state had a very important active role to play, intervening in the economy when private interests threatened the common good. Indeed, just as the serfs had to be emancipated by the state, from above, against the wishes of the majority of the landholding nobility, so the state had a responsibility to intervene and regulate the economy against vested private interests when the common weal was at stake. Ivan Vernadsky saw his role, as a university professor and journalist, to be one of advising the government and helping work out solutions to public problems through free debate in the press.[20] Archival sources indicate that both government officials and revolutionaries alike were rather scornful of the elder Vernadsky at times in his delicate attempts to navigate between the shoals of government repression on the one side and revolutionary action on the other.[21] It was an uncomfortable position his son would later find himself in frequently.

Ivan Vernadsky was called to task by government censors several times for articles that appeared in his journal between 1857 and 1864. One censor, Aleksandr Nikitenko, who kept a diary for this period, described his demeanor in approaching the censors as rather sheepish: "To tell the truth, Vernadsky approached us like a schoolboy; [as if] it wasn't wise to annoy the censorship."

Nonetheless, Vernadsky would not retract what he had published. "After all, in reality," he asked the censor, "what is so terrible about what I have published?"[22] In February 1861, Vernadsky's journal was nearly closed down when the chairman of the censorship committee discovered that the *Economic Index* had begun to discuss openly the necessity for a constitution in Russia. Nikitenko noted that eventually he talked the members of this committee into issuing a severe reprimand instead of banning the journal.[23]

In this period new censorship rules were under discussion, and Vernadsky took the initiative in organizing publishers of various journals in order to influence the legislation. Chernyshevsky wrote to another radical journalist at the time that he participated in one of these discussions at Vernadsky's apartment, specifically the draft of a petition written by Vernadsky to the Minister of Education. But Chernyshevsky, at the time secretly helping prepare a revolutionary appeal to the peasantry and already under close surveillance by the police, believed in the necessity for more rapid and radical change. He quickly grew impatient with Vernadsky's piecemeal and gradualist approach, which consisted, he reported, of "petitioning for improvement in the censorship."[24] Shortly thereafter Chernyshevsky was arrested, and by 1864 Vernadsky himself gave up his struggle with the censorship and closed down his journal.

Earlier in 1862, Ivan had remarried. His second wife, who was a distant relative of his first spouse, was also from a prominent family of Ukrainian nobility.[25] Anna Petrovna Konstantinovich became the mother of Vladimir Vernadsky and the stepmother of Nikolai, a sickly child who was the only issue of the economist's first marriage. Anna Petrovna's interests and personality were quite different from Vernadsky's first wife. Portraits of Anna from this period show a cheerful woman in apparent good health, but she was without the intellectual interests of Maria Shigaeva. Anna Petrovna was a music teacher who had sung in Balakirev's famous choir and who came from a large and boisterous military family of Greek, Serbian, and Ukrainian origin. In contrast to the frail and scholarly Maria, Anna Petrovna filled the house with laughter, music, and a rather stentorian voice. Her father was an artillery general from the time of Nicholas I, a loyal serviceman but a person of narrow interests, without a good education. Anna had four brothers and several sisters. Most of her siblings became military officers or married into such families. Only one acquired a university education. The most prominent brother was Aleksandr Petrovich, who graduated from the General Staff Academy and became the governor-general of several provinces in the 1870s, then a member of the Ministry of Internal Affairs, a bureaucrat and policeman by mentality. Vladimir Vernadsky later described him as "an intelligent, cold man, a careerist." In contrast to the Vernadsky home, Aleksandr Petrovich ran his family like a military encampment. As the scientist later told his son when comparing his mother's and father's families, the Vernadskys had a long scholarly tradition, but "the Konstantinovichs came to education very late."[26] Vladimir Vernadsky obviously felt little rapport with that side of the family. His mother, on the other hand, was by all accounts a warm personality, but very status conscious and not particularly learned.

"My childhood was spent among a mass of relatives," Vladimir later recalled.[27] Anna Petrovna kept a large, open household that was always filled with relatives, servants, and the colleagues and friends of her scholarly husband, a situation not untypical among the nobility and well-off intellectuals from that class. The Russian nobility tended to have large families in this period and to maintain close and intricate ties with numerous relations. Despite his misfortunes with the censorship, Ivan Vernadsky prospered and was able to support such a household in comfort if not luxury. He acquired land in the black-earth region of Tambov province, south of Moscow, and still managed to help poor relations and others needing assistance. He traveled abroad frequently, sometimes with his wife, his children, a nanny, and a tutor as well. Although he prospered financially, his health was delicate. In 1868, during a heated debate in the Free Economic Society over whether large landed estates were detrimental to the welfare of Russian villagers (Vernadsky argued that they were), he suffered a stroke. He recovered but soon resigned his teaching position in St. Petersburg at the Aleksandrovskii Lycee, a closed institute that prepared the children of prominent aristocrats to join the Empire's administrative elite. Vernadsky took a quieter job managing the Kharkov branch of the State Bank in the eastern Ukraine.[28]

Although his father's health was not robust, there is every evidence that the future scientist enjoyed a secure and generally happy childhood. His first clear memories of childhood were not of the capital city, St. Petersburg, where he was born, but rather of the provincial city of Kharkov, where his parents moved when he was five. He also had vivid memories of colorful peasant fairs and the Ukrainian countryside where summers were often spent at the estates of friends and relations. Vladimir was a quiet child, shy around children of his own age, but intellectually curious and precocious. His closest companions were adults and his older half brother, not his two younger sisters. In later years he remembered especially vividly his half brother, Nikolai, who taught him to read and who recited poetry and fairy tales. He felt the talented Nikolai's early death from Bright's disease in 1874 with special regret.[29]

Two others who stood out in his memories of childhood and whose attitudes influenced him were his peasant nanny and his opinionated uncle, Evgraf Korolenko, by then an old man with a full white beard who lived with the family in Kharkov. His nanny, born a serf, was a democratic soul who hated the pretensions of the nobility. She first told Vladimir about life as a serf and about the peasant emancipation in 1861. She was quick to scold him whenever he was rude to her or the other servants: "Who do you think you are? There are no longer any serfs or masters here—we're all people now."[30]

Vladimir's uncle Evgraf loved to talk and argue; in the young Vladimir he found a fascinated audience. As a boy, Uncle Evgraf had met the Tsar, Alexander I, who in the last years of his reign, during his wanderings about South

Russia after the Napoleonic Wars, frequented the household of his grandfather, the Taganrog customs collector. The Tsar promised to enroll Evgraf in the St. Petersburg Corps of Pages, an elite school normally open only to children of high aristocrats. The Korolenkos had not wanted to accept the offer, since it meant separation from the boy, but they were afraid to appear ungrateful to the Tsar. So Evgraf was packed off to St. Petersburg. After Alexander's death in 1825, Evgraf and his teenage friends from the Corps of Pages, hearing about the Decembrist revolt, rushed to Senate Square to join the rebels but were caught and prevented from doing so. Evgraf served in the army during the Russo-Turkish War of 1828, then pursued an undistinguished career in the civil service before retirement. Evgraf was a widely read man, largely self-taught, liberal in his politics and a strong proponent of emancipation for the serfs. The writers of the French Enlightenment were among his favorites, and he kept up with developments in science.[31] Astronomy was a special interest, which he introduced to his young nephew. During long nightly walks in Kharkov, punctuated by an occasional falling star, the old man sparked in his nephew a lifelong interest in understanding the cosmos.

Later, in a letter to his fiancee in 1886, one of a number in which he tried to explain his family background to his future wife, Vernadsky wrote about his memories of Evgraf:

> I recall dark, starlit winter nights. Before sleep, he loved to walk and, when I could, I always walked with him. I loved to look at the sky, the stars. The Milky Way fascinated me and on these evenings I listened as my uncle talked about them. Afterwards, for a long time I couldn't fall asleep. In my fantasies, we wandered together through the endless spaces of the universe. . . . These simple stories had such an immense influence on me that even now it seems I am not freed of them. . . . It sometimes seems to me that I must work not only for myself, but for him, that not only mine, but his life will have been wasted if I accomplish nothing.[32]

In Evgraf, Vladimir Vernadsky sensed that feeling of disappointment and unfulfilled promise that was so common among the nineteenth-century "superfluous men" of the intelligentsia. These were people whom writers such as Turgenev, Chekhov, and others immortalized in literature, individuals frustrated by their own inability to act effectively on the basis of their convictions. Vernadsky's enormous industry and record of accomplishment, in a sense, show a fierce determination to avoid the same fate.

Evgraf had an interesting attitude toward authority, which remained strongly engraved on Vladimir's memory. Talking about Giordano Bruno, Evgraf declared to his nephew that only a madman would allow himself to be burned alive for his convictions. "No, I am like Galileo," he declared. "If the Popes required me, then I would kiss all the crosses twenty times and would not give myself up to be burned!"[33] Evgraf and Vernadsky's father had their own strongly held principles, for which they fought when the season seemed ripe. But in less favorable times, both considered themselves realists in dealing with power. Although Vladimir later indicated that "the cult of the Decembrists was very strong in our house and there was great antipathy toward the autocracy," nonetheless admiration for the Decembrist rebels, who had suffered death or long exile for their actions, was more worship from afar than a principle that led to dangerous risks when the atmosphere for liberal change turned sour. Yet their own sense of frustration with the lack of liberal change, perhaps even disappointment in themselves, took its toll, something that Vladimir sensed in his father and his uncle Evgraf.

If his uncle was one of his favorite companions as a child, his frequently preoccupied father became closer in his teenage years. Grief-stricken after the death of his first son, Ivan Vernadsky turned more to Vladimir for companionship after 1874, generally treating him like an adult and forming a warm intellectual friendship with his only remaining son.[34] As a teenager, Vladimir traveled abroad with his father several summers, and it was during one of these trips, while in Milan, that his father learned about a new decree forbidding the use of the Ukrainian language in publications back home. Angered, he told his son more about the struggles of the Ukrainian people, about their culture, and mentioned one of his wife's uncles, who had been a leader of the illegal Ukrainian Brotherhood of Cyril and Methodius, which sought more freedom for the Ukraine at the time of the reign of Nicholas I.[35]

Curiosity aroused, the thirteen-year-old boy searched through the libraries and used bookstores of St. Petersburg, where the family moved at the end of the summer of 1876. He read everything he could find about Ukrainian history and culture, teaching himself to read Ukrainian in the process. Learning that many books about the Ukraine were in Polish, he also taught himself that language.[36] In later years his knowledge of the Ukraine was to serve him in good stead and led him to champion the founding of the Ukrainian Academy of Sciences in 1918, of which he became the first president. His father's sympathies for other Slavic cultures led the elder Vernadsky back to the literary life of St. Petersburg that fall. The job with the State Bank bored him, and Anna Petrovna disliked the provincial culture of Kharkov. She could talk of nothing but her favorite city, St. Petersburg, where most of her friends and many relatives lived and which was a lively center for her musical interests. Ivan Vernadsky started a bookstore, importing foreign books to St. Petersburg. Attached to it was a small publishing house, which he named the Slavonic Press. Here he not only published books and pamphlets but eventually received permission from the government to print a financial newspaper, the *Stock Market Index*.[37]

Through this business his young son had access to dozens of magazines, foreign and domestic, to which his father subscribed and many books, both permitted and forbidden. Here he was also exposed to the intellectual currents of the 1870s, especially panslavism and populism. His second cousin, the populist writer V. G. Korolenko, who was mentioned earlier, came to work for his father as a proofreader for a time, after being expelled from the Petrovsky

Agricultural Institute for presenting a petition of grievance to the authorities from his fellow students.

To sum up briefly the influence of his family, Vernadsky owed to them his intellectual curiosity and a scholarly tradition that encouraged such curiosity. His liberalism was clearly a family tradition, as was his sympathy for other Slavic peoples like the Ukrainians and his interest in their languages and history. His antipathy toward revolution and willingness to work for change withn the system, which became a distinct trait in his university years, was also a family inheritance. Finally, his own confidence and basic sense of security about himself, however doubtful he may have been about his specific abilities at times, probably came in part from the supportive and secure childhood he enjoyed in a large family that was generally warm in its human relationships and, though perhaps not rich, comfortably well-off by contemporary standards, even for the mass of the nobility. Although his uncle Evgraf had sparked an interest in science, there was no particular indication that he would become a natural scientist before Vladimir entered the later grades of secondary school. A stronger interest in nature came later, in the St. Petersburg gymnasium, and his choice of profession was made only at the university level. Before that time the humanities, especially philosophy, history, languages, and literature, were of greater interest to the young Vernadsky and help account for the philosophical and historical approach he later took in his scientific writings.

In St. Petersburg, Vladimir entered his second year of secondary school, a classical gymnasium, and quickly became acquainted with the lively student subculture of the time, which formed in response to the staid atmosphere of the schools and the repressive educational policies of the state. The contrast between the warmth of family life in a home like the Vernadskys' where individuality and accomplishment were encouraged and the official atmosphere of Tsarist schools, which sought to regiment students for the sake of "public order," could not have been greater. One recent historian has even attributed the growth of the Russian revolutionary movement, which recruited most of its leaders from the student population, to this sharp contrast between the relative permissiveness and warmth children enjoyed in the homes of many nobles and intellectuals and the harsh discipline and intellectual rigidity of the schools.[38] But not everyone responded to this contrast in the same way, and certainly not every student became a revolutionary.

In Vernadsky's case, judging from his own testimony, the atmosphere of the classical gymnasium was important in his decision to become a scientist as well as a cultural reformer. The program in such schools was based on classical languages and literature and largely excluded not only the natural sciences but also modern history, contemporary literature, and philosophy.[39] Although there were a few notable exceptions among the teachers, Vernadsky later recalled that the "chief misfortune was the deadening spirit of teaching."[40] Even the good teachers who knew Greek, Latin, and classical civilization well could not transmit their knowledge effectively. They were forced to follow a rigid curriculum prescribed by the Ministry of Education under the counterreformer Count Dmitrii Tolstoi. Many of the teachers were foreigners who were thought to be politically more reliable but were frequently unable to speak Russian well and sometimes acted as spies for the administration. As Vernadsky put it later, such teachers were generally "cut off from Russian life and foreign to the interests of our country, carrying out conscientiously the anti-national official program of the government."[41]

Many students compensated for the deadening atmosphere of the school by pursuing their own interests in small circles of friends. The rather shy thirteen-year-old who had just arrived from Kharkov was befriended by the leader of one of these groups, A. N. Krasnov, a cheerful, oval-faced lad who was to become a prominent botanist and plant geographer and who remained a lifelong friend. They became especially close when Krasnov came down with scarlet fever and Vernadsky sent him books and letters. Krasnov was the son of a Cossack general from the Don region and already considered himself an ardent panslav and enemy of German influence in the Russian government and Eastern Europe. The German minority in Russia, particularly the nobles from the Baltic region, held a large number of high positions in the military and civil service and played a major role in the Imperial Academy of Sciences. The Cossack connection and panslavist sympathies of the two boys helped form a common bond, particularly as their friendship blossomed during the Russo-Turkish War of 1877–78. With their schoolmates they discussed every aspect of the campaigns in that war.[42] Vernadsky's schoolboy diary indicates his strong sympathies at the time for the liberation movement of the Balkan Slavs against the Turks, a movement which was reluctantly supported by the Russian government. In other respects the boys were different, however. Krasnov at this time was more interested in the world of nature, in plants and animals, especially insects. He led his friends on expeditions through the environs of St. Petersburg to collect specimens and at home built a terrarium that fascinated the others boys.[43]

"For me," Vernadsky later recalled, "these activities were new and more than the gymnasium of that time taught me—for my interests then were in other fields, in the areas of philosophy, history, geography, religion, and Slavic languages."[44] The boys not only discussed science and gathered specimens but began to try chemistry experiments, which sometimes ended in accidents that alarmed the boys' parents and neighbors. "Strangely, it was the demoralizing classical gymnasium of Tolstoi that stimulated my interest in natural science thanks to that hidden, internal, underground life which thrived in such circumstances. . . ." Krasnov opened up for Vernadsky a lively new world, "before which the dry and stifling instruction of the official school paled completely."[45]

The friends formed a circle and began to publish a small, handwritten journal on scientific themes, richly illustrated with their own drawings. At first largely concerned with insects, the journal gradually broadened its focus to include other scientific themes and contemporary literature, forbidden and otherwise.

registered in history-philology. By 1888, the proportion had declined to 7 percent, and by 1899, only 3.9 percent of all Russian university students went into this area.

Though not all these changes had been instituted when Vernadsky matriculated in the fall of 1881, he entered during a period of transition from a freer university atmosphere to one of tighter controls and greater tension. Vernadsky himself hesitated for a time between the History-Philology Faculty and the Mathematics–Natural Science Faculty.[53] The other two faculties at St. Petersburg University were the medical and the juridical. It was his friendship with Krasnov, who entered the Mathematics–Natural Science Faculty, as well as the brilliant reputation of that faculty which were among the determining factors. The greater political vulnerability of the History-Philology Faculty and a strong distaste for the classical gymnasium were also considerations. In fact, as Vernadsky noted, more of their fellow graduates from the gymnasium entered the natural sciences, hoping for more freedom of inquiry in these fields. Vernadsky later explained: "The university had an enormous significance for us. It first provided a free outlet for that underground life which had flourished in the gymnasium."[54] Perhaps never again, his friend Krasnov emphasized, would there be such an opportunity to work with a collection of scientists as brilliant as Mendeleev, Butlerov, Beketov, Sechenov, Menshutkin, and Dokuchaev, among others who were professors in the Mathematics–Natural Science Faculty.

Russian universities at the time Vernadsky entered trained much of the intellectual elite, but not the country's administrative elite. This division within the Russian elite, and the wide gap in mentality that resulted, account in part for the suspicion with which Russia's bureaucratic rulers viewed the universities. The largest single group of high administrators in the Tsarist government during the late nineteenth century were graduates of a few institutes which charged higher tuition than the universities and were closed to all but the children of the upper aristocracy. The Corps of Pages, the Aleksandrovsky Lycee, and the Institute of Law were three of the most important of these institutions. For example, of eighty-four persons in 1897 who were government ministers, department heads in the government, of members of the State Council, only twenty-two were university trained, and all but two of the latter were from St. Petersburg or Moscow Universities. Thirty-seven were graduates of the three closed schools listed above, and fifteen had received military educations.[55]

The universities, therefore, prepared persons generally for a different track in life not closely associated with the ruling elite, primarily for teaching, the free professions, and middle-level government positions, and trained them in the skeptical, critical spirit of nineteenth-century secular culture. They also drew the majority of their students from a different constituency. The large majority of university students in the late nineteenth century were not well off. Most were the children of the economically declining nobility, of teachers, parish priests, and a growing contingent from the lower middle class of artisans, small shopkeepers, clerks, and so on. Children of the small Russian bourgeoisie (upper middle class) frequently did not attend the universities but preferred the more specialized technical and commercial institutes run by the Ministry of Finance and other technically oriented branches of government. This left the universities in the late nineteenth century with only a minority of students in higher education, a minority that was socially and economically very insecure. Two-thirds of the university students in the 1870s needed financial assistance, and only about one-third of those who matriculated eventually were able to graduate.[56] Although by European standards of the day, the Russian government provided a large amount of financial assistance, it was far from adequate to meet even the basic needs of most university students. A spartan existence was the rule for a large majority of these young people.

Vernadsky and his closest friends at St. Petersburg University were exceptions, then, in coming from families that were financially comfortable. Vernadsky and a number of his friends continued to live at home, and this also set them apart from poorer and more provincial students who flocked to St. Petersburg University from many other sections of the country, cutting off closer ties with their families in the process. Vernadsky and his friends belonged to a small group of university students who consciously tried to set themselves off both from the so-called white gloves on the right (the tiny minority of conservative and aristocratic students who attended the university) and the vast majority of students on the left, often radicals who were usually poorer and often more alienated from their families and parental values. At first viewed rather negatively by radical students and called "mama's boys" and "little barons,"[57] Vernadsky and his friends won wide if sometimes grudging respect for their integrity, seriousness of purpose, and emphasis on serving moral ends. Unlike so many students who made radical activities an end in themselves, so-called *kulturniki* ("culturals"), like Vernadsky, were very serious about attending classes and receiving a good education. Their professed goal was to serve Russian society by raising its scientific and cultural level. For the kulturniki, their university years became the crucible out of which they tried to forge a "middle way" distinct both from the narrow careerism they criticized among more conservative and apolitical students and the revolutionary impatience they considered so wasteful and superficial among the majority of more radical students.

Even in their manner of dress the kulturniki usually stood out from the mass of students by their neatness. Many radical students in the early 1880s, before uniforms became mandatory for all university students, liked to affect the dress and manners of the Nihilists of the 1860s—red peasant-style blouses, high leather boots, and wire-rimmed glasses. As a former radical student at St. Petersburg University in this period, V. A. Posse, later a well-known Marxist journalist, wrote: "I became a part of the revolutionary-minded students. Students of this element, not only by their views and mood, but by their appearance, could be distinguished from the kulturniki. The revolutionaries dressed more carelessly, behaved more freely and laughed at the correctness and loyalty of the 'culturals.' However, there was no real enmity between these two elements but, on the contrary a kind of osmosis developed between them."[58] By osmosis, Posse meant a continual interchange of views, a certain sympathy and sometimes cooperation between the two groups. This same writer remembered Vernadsky as a good natured person, "always with a kindly smile . . . very soft in appearance but very determined once he had set himself a goal. It seems to me that Vernadsky, as well as Sergei Oldenburg, (one of Vernadsky's closest friends) had already set themselves the goal not only of becoming professors but also members of the Academy of Sciences."[59]

Three instructors caught Vernadsky's interest. The first was D. I. Mendeleev, inventor of the periodic table of elements, who in 1880 had attracted wide sympathy in the Russian intelligentsia when he was rejected for membership in the Imperial Academy of Sciences by one vote. A wave of nationalistic indignation swept through the Russophile sectors of educated society, who attributed his rejection to the strong German contingent in the Academy. Actually his rejection was probably due to the machinations of the Russian minister of education, Dmitrii Tolstoi, who was also president of the Academy and feared strengthening a contingent in the Academy who were critical of the government, as well as to interfactional tensions between natural scientists and humanists who feared domination of the Academy by those like Mendeleev who were advocates of the exact sciences.[60] Such conflicts crossed ethnic lines and had little to do with nationalistic prejudices, but it is significant how many persons, including Vernadsky and Krasnov, interpreted his rejection in a nationalistic way. The irony of his abortive election to the Academy is that Mendeleev was politically moderate, even rather conservative, a monarchist at heart who believed strongly in the need for industrialization and strengthening of the natural sciences and technology. Such beliefs made him highly critical of current government policies in culture and education. In his university lectures he criticized the dead hand of the classical curriculum in secondary schools and emphasized the importance of the exact sciences and their application in improving the national economy. Vernadsky indicated that Mendeleev's lecture hall was always filled to overflowing: "We entered a new and wondrous world during his lectures, as if released from the grip of a powerful vise."[61]

Another of Vernadsky's teachers was also at the forefront of modern chemistry. Butlerov, one of the pioneers of modern structural chemistry, became involved in a dispute with Mendeleev during Vernadsky's student years over the nature of the atom. Mendeleev argued that the atom was indivisible, the smallest possible unit of matter, believing that only this view was compatible with his periodic table of elements. More than a decade before Becquerel's discovery of radioactivity in France in 1896 and the revolution in physics that followed, Butlerov was considering the possibility that the atom was divisible and might be more complex than Mendeleev and most other chemists of the time believed. Vernadsky once ran across Butlerov in the halls of the university, good naturedly expounding on this view to a group of students who disagreed strongly. Butlerov argued, "we consider that atoms are indivisible but this means that they are indivisible only by means that are presently known and accessible to us; and they remain stable only in those chemical processes we now know. It is possible that they are divisible under new processes which we may discover."[62] By his third year, Vernadsky had decided to specialize in crystallography and mineralogy and was taking Mendeleev's course in inorganic chemistry at the time. Asked for his opinion by Butlerov, Vernadsky defended the orthodox view. "Unfortunately, we have no experimental material that would make us doubt the indivisibility of the atom," he remembered arguing, citing the authority of Mendeleev. "Of course, experiments are necessary," Butlerov replied and referred to experiments under way in his own laboratory. "And as to authorities," he added in good humor, "I can cite the authority of the famous Arago [a French astronomer and physicist of the early nineteenth century]. Do you know, gentlemen, what he constantly told his students? Unwise is the one who, outside the field of pure mathematics, denies the possibility of anything!"[63] Vernadsky recalled his encounter vividly in later years when defending the tentative nature of all truths, whether in natural science or social theory. The relative nature of knowledge and the need for tolerance of opposing viewpoints were to become cornerstones of Vernadsky's scientific liberalism. By the early twentieth century most of Vernadsky's own work was directed no longer by Mendeleev's dogmatic defense of an old truth— the indivisible atom—but by the revolutionary new concepts of atomic structure coming out of Western Europe, including the laboratory of the Curies in Paris, where he was to work during the early 1920s.

The third and most influential of his teachers was V. V. Dokuchaev, who held the chair in mineralogy at St. Petersburg University and became the director of Vernadsky's first independent research in science. In the years when Vernadsky was a student, Dokuchaev was at the height of his intellectual powers and was involved in developing a new interdisciplinary field, creating an original school of Russian soil scientists.[64] The son of a village priest, Dokuchaev knew traditional Russia well and to the end of his life his scientific interests were focused on improving the conditions of life for Russia's rural population. Trained as a geologist and chemist, Dokuchaev was especially interested in soil exhaustion, one of the most important economic problems facing Imperial Russia. Building on the work of the German scientists Liebig and Sprengl, Dokuchaev went beyond them to understand not only the chemical composition of soils but how they developed in interaction with their environment, how and why they varied regionally. One of his most ambitious goals involved regional expeditions to create a series of soil maps of Imperial Russian territory. This work was supported financially by several of the most prestigious scientific societies of the empire, the Free Economic Society, in which Vernadsky's father had also once been active, and the St. Petersburg Society of Naturalists, which was centered at the university and drew into its activities many of the brightest students in that city.[65]

A brilliant, at times temperamental, man who to an acute observer already showed signs of the intensity and mental strain that brought his life to a tragic end in 1903, Dokuchaev in these years attracted a large following among students, many of whom, including Vernadsky, participated in his scientific expeditions to the black-earth region of rural Russia. Vernadsky, in a tribute to his deceased mentor published in 1904, characterized him as a "self-made man . . . a real Russian of natural gifts who went his own way, who was formed entirely in Russia and was completely foreign to the West which he did not know—just as he knew no foreign languages—and which he visited only toward the end of his life."[66] Vernadsky's characterization somewhat understates Dokuchaev's debt to German science, but unlike most Russian professors of his day, he did not receive any of his higher education in Western Europe and was not particularly interested in vying with Western scientists for priority and scientific fame. Dokuchaev's nationalism and moral attitude of using science to improve Russian society no doubt impressed Vernadsky as well as many other students and helped to account for his popularity.

Beyond that, Dokuchaev was an unusually gifted teacher, a refreshing contrast to so many of the instructors these students had known in secondary school. As another of Dokuchaev's students explained, "young people were drawn to Dokuchaev's lectures. . . . He spoke without pathos or gestures, without any oratorical flourishes, but quietly and clearly, concisely, with precision and to the point, providing examples throughout. . . . Of all my teachers I know only one other with such a gift for convincing—D. I. Mendeleev."[67] When Vernadsky first began to study with Dokuchaev in 1883, the soil scientist was only thirty-seven and had just completed his first major book, *The Russian Black-earth Region*, which was the culmination of seven years' intense work.[68]

Just as Dokuchaev viewed soils as complex phenomena, partly inorganic, partly organic, developing slowly over time through the interaction of plants and animals, rocks, climate, etc., so he viewed all of nature similarly, as a complex, interacting whole. Dokuchaev reinforced Vernadsky's interest in a holistic and historical approach to science, to a study of how the phenomena of nature have changed over time. Recalling his student years, Vernadsky wrote in 1935:

> While reading mineralogy at the university in St. Petersburg, I began on a path at that time unaccustomed. This was in connection with the work and contact during my student years and immediately afterward (1883–97) with the great Russian scientist V. V. Dokuchaev. He first turned my attention to the dynamic side of mineralogy, the study of minerals through time. . . . This defined the whole course of my teaching and study of mineralogy and was reflected in my thought and the scientific work of students and colleagues.[69]

A synthetic and interdisciplinary approach to science was also to become a hallmark of Vernadsky's career over a period of sixty years. In this approach he was strongly influenced by the legacy of his mentor. As Dokuchaev summed it up in 1898–99:

> Looking more attentively at the great discoveries of human knowledge, discoveries, one might say, which have revolutionized our view of nature from top to bottom, especially after the work of Lavoisier, Lyell, Darwin, Helmholtz, and others, it is impossible not to notice one very real and important shortcoming. . . . They have studied chiefly separate bodies—minerals, mining deposits, plants and animals, and individual phenomena—fire (vulcanism), water, earth, air, in which, I repeat, science has achieved astonishing results, but not their interrelationships, not that genetic, eternal, and always orderly link which exists between forces, bodies and phenomena, between inert and living nature, between plant, animal, and mineral kingdoms on the one hand and man, his daily life and even spiritual world, on the other. But it is expressly these interrelationships, these lawful interactions that comprise the essence of a knowledge of nature, the core of a true natural philosophy—the best and highest achievement of scientific knowledge.[70]

Under Dokuchaev's direction, Vernadsky completed his first original scientific work. In 1882, Dokuchaev had been asked by the local government (zemstvo) of Nizhegorodskii province along the Volga River below Moscow to do a survey of the quality, types, and distribution of soils in that province. Dokuchaev invited his students to accompany him on expeditions to this area, and Vernadsky frequently did so. Vernadsky kept a detailed journal on these trips and noted his observations not only of the soil but also of many aspects of the natural environment, including the destruction of forest lands and the way the Volga had changed its course and affected the landscape.[71]

In one of the notebooks Vernadsky kept of the expedition, we find a hint of an idea which formed the core of much of his early scientific accomplishment. Just as Dokuchaev was trying to explain why soil types varied from area to area, Vernadsky began to ask the same question about minerals. "Who knows," he wrote during one of these student expeditions along the Volga, "perhaps there are laws for the distribution of minerals, just as there are reasons accounting for the possibility of forming one or another chemical reaction precisely in one place and not in some other."[72] Questions such as this led him to investigate why certain minerals were found together in particular regions and what conditions, including chemical reactions in past geological time, created deposits of a particular type in one area and not another. Dokuchaev strongly encouraged Vernadsky's interest in the genesis of minerals. Most mineralogists of the period were not interested in a genetic approach to their discipline but were concerned with a static description of the qualities and composition of minerals. "This can be the program for a lifetime of work," Dokuchaev told his pupil, "and it is worthy of it."[73] As it turned out, however, Vernadsky was too restless an intellect to confine himelf solely to this topic and explored many other aspects of science before his career ended some sixty years later.

During his student years, Vernadsky published two scientific articles. The first was a description of an excavation near the city of Nizhnii Novgorod, in the province where Dokuchaev was working, and an analysis of its minerals.[74] The second consisted of observations on the dwellings of a type of prairie rodent.[75] Although not training to become a biologist, Vernadsky retained his schoolboy interest in animals, sparked by Krasnov's circle. It is significant that his debut in the world of scientific publications was inaugurated by one article on the inanimate world and another on the animate. The interrelationship between what his mentor called "inert and living matter" was not yet the dominant concern it would become in his scientific work after 1916, but Dokuchaev had already taught him to ask questions that would eventually lead his mind in that direction.

The subject of Vernadsky's undergraduate thesis remained within the confines of the inanimate world. Dokuchaev proposed the topic: "On the Physical Properties of Isomorphic Mixtures." In 1885 Vernadsky completed his work on isomorphism—the ability of a series of elements to replace one another in minerals—so well that his instructors proposed that he remain in the university to prepare for a career as a university professor.[76]

For Vernadsky, the university was not only a place where he could acquire the best scientific culture of the time but also a place where ideals and moral values were actively discussed and tested in the lively student culture of the time. In this sense, what happened outside the classroom and beyond the framework of formal instruction was at least as important as his formal education in forging a sense of direction and future goals. Vernadsky joined a series of circles and student organizations whose members concerned themselves with the so-called accursed questions—about the meaning of life, about personal ethics and responsibility, the immortality of the soul, etc. Such questions, to a more hedonistic, cynical, or careerist general might seem unfashionable, but Russian university students of Vernadsky's generation took themselves very seriously, something perhaps inevitable when their own role as students was continually questioned by so many of their contemporaries. Russian university students lived in a very hostile environment. Those who valued a university education formed a thin layer on Russian society in the 1880s. They were surrounded by suspicious government officials who sent soldiers and footpads into the university at the slightest provocation and recruited janitors and doormen to spy on the student population in the interim.[77] Local tradesmen, like the butchers and other shopkeepers and artisans, generally disliked the students and considered them traditional enemies in the large cities. Such tradesmen later formed the core of the Black Hundred gangs who beat up students, Jews, and intellectuals during the revolutionary troubles in the early twentieth century.[78] In such an environment, students indulged in a great deal of introspection and self-justification and needed mutual support and reinforcement to persist in their educational goals. Friendships tended to flow deeply and were often much closer in such hostile settings than they might otherwise have been.

Vernadsky joined several student groups, the most important being that of the Oldenburg brothers, Sergei and Fedor. The members of this circle (kruzhok), for the most part sons of the Tsarist generals and civil servants and numbering between ten and twelve, eventually including their fiancees as well, set themselves the conscious task of developing a common world view and a common set of goals in life.[79] Although there is no reason to believe that they fully succeeded in this task, for Vernadsky the friendships formed in this circle not only lasted many decades, they were a strong support and stimulant for much of his later scientific career and work as a public figure. Although the Oldenburg group differed on many questions, what they shared in common was a belief in modern knowledge and its transforming power and in the responsibility of the educated to work energetically, through peaceful means, to spread their knowledge in order to improve the material and spiritual conditions of Russian life. It was for this common set of beliefs, which linked them to Lavrov's brand of populism in the past (the movement "to the people" of the early 1870s) and in the future to the Russian liberal movement of the early twentieth century, that the Oldenburg group provided an important leadership contingent.

In fact, the distinction between Russian liberalism and Russian populism is, in part, not so sharp as most Westerners might expect. Many of the leaders of Russian liberalism, including Vernadsky, were strongly influenced by populist ideals and should perhaps be viewed in part at least as the successors of the evolutionary wing of populism, just as the Socialist Revolutionary party of the early 1900s became heir to the more revolutionary wing of the 1870s populist movement.[80] There were other important differences as well, of course. The liberals had a negative view of the peasant commune and of socialism as a goal, but they shared populism's belief in science and in the need to raise the cultural level of the people. There is little evidence in Vernadsky's own writings or private papers that the traditional goals of Western *laissez-faire* liberalism played any significant role in his world view. The defense of private property as an indispensable guarantee of liberty was, for example, not something his writings were concerned with. In fact, Vernadsky generally put the commonweal above the right of private property. For example, in 1884, while still a student, Vernadsky wrote in his diary about the destruction of forests in the Volga region, which he witnessed while on expedition with Dokuchaev. Concerned with the influence of forests on climate and the effects of erosion which followed the destructive lumbering practices he witnessed, Vernadsky wrote:

> Here they are cutting down the forests; for the most part these lands belong to the nobles (Sheremetevs, Golitsyns, Matveevskys) and to the state; the peasants own very little. Only the state-owned forests are being preserved; the

nobility and merchants are cutting down their forests mercilessly . . . it is necessary to use these lands with reason and good sense; otherwise this is theft, the stealing of the wealth of all; it is necessary to use the force of law against those who employ exploitative practices, such as the Matveevskys, Sheremetevs, etc.[81]

Several of Vernadsky's other diary entries for 1884 provide clues about how his own world view developed in heated debates among his friends and fellow students. Vernadsky noted that one of his friends (not a member of the Oldenburg group) had argued that "the scientist by his work alone brings value to society and therefore does not need to know, must even 'forget' about the society he is helping." Vernadsky added, "I hold a completely different opinion."[82] He then tried to explain the basis of the difference. For him, "the task of a person is to achieve the most good for those around him." This was not a task justified by an afterlife, a desire for personal immortality, or belief in a supreme being outside nature who would judge people for their acts. Vernadsky noted in his diary that he did not believe in any of these things at the time. Rather, the "life of personal saintliness" which he set as a goal in these years was a secular saintliness.[83] Since he thought that his personal existence would end with his death, his goal was to achieve the greatest possible happiness in this mortal existence. He felt he could achieve such happiness only by helping those around him, by achieving greater understanding of nature and people through science and culture. Material well being but not luxury or wastefulness was necessary as a means to these higher ends; but for him, real happiness could only be achieved by creating and sharing culture. "Placing as a goal the development of mankind, we see that this is achieved by a variety of means and one of these is science."[84] It was not enough simply to be a good professional scientist narrowly confined within a particular specialty and unconcerned with science in general or overall social conditions.

For Vernadsky, human progress was not inevitable. He was not a facile believer in the onward advance of civilization as something guaranteed by the laws of history. "This is something that either may be or may not be . . . the conditions permitting scientific activity may be destroyed; everything that happens in the government and society, sooner or later depends on us. This conclusion leads to the necessity to be active in government and society, to struggle for your ideal so that you and those who come after you might achieve the greatest possible happiness."[85] Struggle and the need to gain power to achieve one's ideal, therefore, became essential to Vernadsky's pursuit of happiness. In a diary entry made in June of 1884, he defined once again two areas as essential to his own happiness: the development of science (*nauka*), which would give him pleasure by learning more than was known prior to his existence, and the development of humanity, the pleasure in struggling to put into practice his ideals. To achieve these goals, Vernadsky felt he needed not only to improve his mind but his character—his openness, willingness to speak out and defend his opinion, his independence, and ability to act on his convictions.[86]

Until his university years, Vernadsky had always felt himself uncomfortable among strangers, dependent on his family and overly shy. His activities in student groups as well as participation in Dokuchaev's expeditions became important to his own sense of self-development.

Some of these activities got him into trouble with the authorities as well as with his family. It was during one of these events that Vernadsky first met the Oldenburg brothers and their friends. In November 1882, a group of radical students called for a student mass meeting (skhodka) to consider actions of the government which they considered insulting to the student population. The skhodka was a well-established tradition, a kind of student rally which one historian has described as "derived from the folk habits of the Russian peasantry transferred to the bureaucratic setting of the Western school."[87] They were called periodically to discuss grievances students might have with the government, faculty, or university administration and were an integral part of the radical student subculture in Russian universities. Skhodki were considered illegal by the government and disruptive by the faculty, but most students defended the institution as one of their few outlets for protest. In 1882, the faculty of St. Petersburg University were particularly concerned not to present the government with any pretext for further interference in university life and their attitude was shared by some of the kulturniki among the students. When the Oldenburg brothers and their friends heard about the preparations for a skhodka and the radicals' plan to use this meeting as a launching point for a street demonstration outside the walls of the university, they opposed it, fearing disruption of the university. When the skhodka convened, however, they attended, speaking against any provocative acts as a threat to academic freedom, since they would only play into the hands of government reactionaries who wanted a more rigid university statute, removing what semblance of autonomy still remained to Russian universities. The Oldenburg group, like most of the faculty, called for a university dominated neither by the political right nor the political left, but rather a place where learning and freedom of thought and inquiry could flourish.[88] In doing so, of course, they were making a political statement of their own.

Before the student skhodka of November 1882 could decide on any action, the university was surrounded by police and armed troops. Students who were not politically involved were invited to leave the university grounds. However, the Oldenburg group and other non-radical students like Vernadsky, angered by outside intervention, decided to remain as a sign of solidarity against this show of force by the government. They were incarcerated for a time and interrogated, then most of the students were released. However, a number of the radical leaders were arrested, expelled from the university, and exiled. The Oldenburg circle, which had originally opposed their plans, ended up raising money to help the radical leaders after their expulsion. As a result of this encounter, Vernadsky, Krasnov, and other friends from the natural science faculty became friendly with members of the Oldenburg circle, who were mostly enrolled in

History-Philology and the Juridical Faculty.[89] They shared similar attitudes both about student radicalism and the repressiveness of the authorities. As another member put it, "these events knit us much more closely together, uniting our circle and enriching it with several new members."[90] One of the first actions of this circle, after Vernadsky joined it, was to set up a study group to report on the history and organization of universities and student life in Germany, Italy, England, and the United States, with the hope that they could learn from foreign experience.[91]

Two members of the Oldenburg circle suffered unpleasant consequences for their participation in the student unrest of November 1882. Sergei Kryzhanovsky, the son of a school master from Warsaw who in later years left the Oldenburg group and became a high official in the Ministry of the Interior, was detained by the police for several days when he answered, in reply to a question about who had caused the skhodka: "The police and the government administration." Prince Dmitrii I. Shakhovskoi, who soon became one of Vernadsky's closest friends, suffered an even more serious penalty from the university authorities. Despite his anti-radical position, they condemned any active participation in organized student activity and to make an example, sentenced him to five days in a university prison. The University of St. Petersburg came equipped with a student cellblock, or *kartser*. Although the incarceration was a black mark on his record, his actual imprisonment was very lenient. His friends were allowed to visit and when he pleaded that he was needed at his sister's wedding, which was scheduled to take place during his imprisonment, he was allowed to shave, don a frock coat and ride to the Anichkov Palace for the wedding. It seems that Dmitrii's sister, the Princess Nataliia Ivanovna Shakhovskaia, had as her "sponsoring parents" at the wedding the Emperor Alexander III and his wife, Maria Fedorovna, who was a close friend of the Princess.[92] Prince Dmitrii's family came not from the old aristocracy but from a family with a mixed history. His father was a general in an elite guard regiment and was descended from Prince Shcherbatov, one of Catherine the Great's severest critics among the eighteenth century nobility. His grandfather on the paternal side had been a Decembrist who had died in exile in Siberia. Dmitrii remembered the bitterness in his own family about the destruction of this idealistic young man who perished without ever knowing his own son, Dmitrii's father.[93] Prince Dmitrii himself was not discouraged from activism by his temporary jailing and later became active in the zemstvo movement and was one of the founders of the Constitutional Democratic party. Shakhovskoi was to die, stripped of his estates and almost forgotten, in the Soviet Union around 1940; his daughter Anna Dmitrievna, became Vernadsky's private secretary in the 1930s and was his literary executrix and curator of the Vernadsky Museum in the Soviet Academy of Sciences until her own death in 1956.[94]

Although Vernadsky was not jailed, this encounter with the authorities in November of 1882 proved to have an unpleasant consequence for Vladimir at home. The student protest had occurred during his father's last illness, and when Vladimir's mother learned that he had attended the skhodka and had been detained by the police, she caused a scene. As Vernadsky's Soviet biographer later noted, Anna Petrovna had never shared Vladimir's political or scientific interests. She disliked intensely the white mice Vladimir kept at home and even the aquarium and canary he kept in his room because she thought they made the house less tidy. "So this is what your father's politics have led to," he remembered her berating him in a whisper, so that his bedridden father would not hear, adding: "Do you want to end up in Siberia, like Chernyshevsky?" She accused Vladimir of not thinking about her or his father, powerful inducements for caution to a son who had heretofore apparently been a model of filial devotion. [95] The strain between his mother's values and his own continued through his university years, and although he lived at home until his marriage in 1886, the relationship between mother and son seems to have been marked by tension and perhaps even affected his choice of a bride.

The tension Vernadsky felt between his mother's values and his own probably accounts in part for the portrait of him by another member of the Oldenburg circle, I. M. Grevs, the son of a retired Tsarist officer and landowner, who later became a noted literary historian and biographer of Turgenev. Grevs, who greatly admired Vernadsky for his industry and scientific accomplishments, nonetheless sensed a certain nervousness and reserve in the large, bearded student who looked a bit like a friendly bear in his photographs from this period. "Sturdy, with a rosy complexion, reservedly friendly and smiling, he looked at people with an even-tempered, critical attention. But in his movements one felt a nervousness; a light-hearted mockery crept into his words and tone and in his actions, sometimes a kind of youthful exuberance. . . . Vernadsky was a clearly expressed individualist but never anti-social."[96] Whatever his inner tensions and contradictions, Vernadsky flourished in the atmosphere of his new-found friends. The Oldenburg circle was dedicated to the ideal of "the self-development of the individual personality" as well as to serving people through cultural enlightenment. [97] In fact, they saw no contradiction between the two ideals but rather a necessary linkage: the individual could develop his talents only through service to others. The Oldenburg circle provided a lively outlet for Vernadsky's values, therefore, in a way he was not likely to find in a home now dominated by his mother's anxieties. Fedor Oldenburg later expressed the common values of the group in the following shorthand: to work and produce as much as possible, to use (consume) as little as possible, to treat the needs of others as if they were one's own. The intended tone of asceticism was, in part, influenced by the spiritual crisis of the famous novelist, Lev Tolstoi, which occurred in these years. Tolstoi's crisis was followed with great interest by members of the Oldenburg circle, including Vernadsky, who read his moralizing tracts as soon as they appeared and shared Tolstoi's interest in simplifying material wants and enriching their spiritual life by drawing closer to the Russian people. [98] The tone of asceticism was in sharp contrast to the life of luxury, or at least the very comfortable existence led by most of their families.

The Oldenburg circle met once a week, usually on Thursdays and usually in the Oldenburg home, where Fedor and Sergei lived with their mother, the widow of a Tsarist general descended from the Mecklenburg nobility of Germany.[99] The Oldenburgs had emigrated to join the Russian service in the early eighteenth century under Peter the Great. The family was thoroughly Russified by the late nineteenth century, when the father of Fedor and Sergei retired from the Tsarist service and moved from Warsaw where he had been stationed as part of the Russian occupation forces, to St. Petersburg, in order to supervise the education of his two sons. His death had left Fedor and Sergei responsible for their own educations. The circle they formed, at first composed of boyhood friends from Warsaw who had enrolled at the university in St. Petersburg, became an inseparable part of that education.

Weekly meetings of the Oldenburg circle often lasted long after midnight, sometimes till morning, during which time circle members discussed a variety of topics, including Russian novelists from Dostoyevsky to Uspensky, philosophical and religious questions, future careers and goals, as well as contemporary political and social issues. During one of these marathon sessions, someone suggested that the group pool their resources and purchase a small estate. The estate could be used as a retreat by any of their members in time of need and would provide a periodic gathering place for reunions, so that members would not become permanently separated after their graduation from the university. Although the estate, to be called "The Haven," was never purchased, members of the circle frequently did spend part of their summers together on one or another country estate and ever afterwards referred to themselves as the "Haven Brotherhood" (*Bratstvo Priiutino*). Whenever they were able, members also gathered the evening before New Year's Eve, and corresponded frequently, following each other's careers with great interest and helping when they could.[100]

Unable to meet with the group his first year after graduation, Fedor Oldenburg, who had just moved to the provinces in order to direct a girls' school, wrote on New Year's Eve 1885 about the meaning this circle had for his undergraduate years:

> Only two hours remain before the New Year and I wanted to dedicate this letter to the other members of our circle, with whom my best memories of student life, perhaps of my entire life are linked. It seems to me that I am much indebted to our circle, much more than to any lectures, more perhaps than to all the rest of the circumstances of my life. How many spontaneous, warm and joyful meetings it provided and how much it did for our development![101]

In later years, Vernadsky painted the circle in a less emotional but equally favorable light. Writing a memoir about his friend Krasnov in 1916, he explained the significance of the Oldenburg circle as part of a larger tendency in Russian student life:

In the beginning of the 1880s, along with purely socialist moods, there existed other tendencies, close to the latter but not included in it boundaries. The purely socialist tendency was permeated by a feeling of social morality, close in its philosophical ideals to scientific positivism, linked with a negative attitude to religion, art, and especially to [contemporary] political life. These other [non-socialist] tendencies were not different in their moral and democratic aims and were in agreement with the necessity to work for the impoverished, for the mass of the people; they also rejected the historically given order of things. But they did not share [with the socialists] the same attitude toward religion, art, philosophy, political life, or science which was part of the socialist mood of youth at that time. Many intellectuals considered it difficult to reconcile socialism with other sides of the human spirit that were dear to them—with a feeling for their nation or the state, and even so with their belief in the freedom of personality. They could not accept socialism on faith or as a scientifically proven truth, as it was accepted by its proponents. . . . These tendencies were rather formless, but in a period marked by the outbreak of terrorism by the People's Will Organization (a socialist group) and about the time when Russian Social Democracy began to form, circles also began to form out of these other tendencies. Some led to Tolstoianism, others created the cadres of the democratic zemstvo movement, leading finally to the Union of Liberation and the Constitutional Democratic party [after 1905].[102]

The Oldenburg circle, Vernadsky went on to explain, was part of the latter tendency and formed an important seedbed of Russian liberalism. Although its members were generally attracted by Tolstoi's desire for simplicity in material things, they differed from Tolstoi in considering knowledge "one of the highest achievements of the human spirit. They acknowledged the value of contemporary culture in general and also acknowledged the necessity for law courts and the state." Tolstoi, they believed, undervalued secular knowledge and was something of an anarchist with regard to the state. Unlike Tolstoi, they considered political reforms essential.[103] Their members, while sometimes cooperating with the Tolstoians, for example, in providing relief during the famine of 1891, found their chief outlets in education, research, local self-government (the zemstvos), and eventually in the democratic liberalism of the Kadet party.

As students in the 1880s the two chief outlets for members of the Oldenburg circle, besides their studies, were the Literary-Scientific Society of St. Petersburg University, which lasted from 1882 until closed down by the government in 1887, and a study group associated with the St. Petersburg Committee on Literacy, an organization with populist ties whose activities were aimed at bringing adult education to the working class of the capital. The Literary-Scientific Society formed a brief interlude in Russian university life in this period, an exception to the government's prohibition against legal corporate life or student organizations. In 1881, a small group of right wing students petitioned to form such a society, which would ostensibly be apolitical. Its purpose would be to arrange lectures and discussions on literary and scientific themes. The organization's secret purpose was to form a group of students loyal to the autocracy and to disseminate monarchist views in a struggle against the revolutionary and constitutionalist moods among the majority of students. Chartered by the government late in 1881, by early 1882 some of its conservative student members began to brag about its true purpose and the advantages of membership. Supposedly its members would receive comfortable government positions after graduation.

The Literary-Scientific Society's faculty sponsor, Professor Orest Miller, a specialist in Russian literature, had been unaware of the ulterior motive of the student founders; and when the word slipped out, he threatened to resign. However, he reconsidered when the Oldenburg group rallied to his defense and gained control of the society from the conservatives. The Oldenburg circle turned the Society into a lively center of intellectual life where a variety of ideas and viewpoints could be discussed in a spirit of mutual respect and tolerance.[104] Tolerance for opposing viewpoints was a rare quality in Russian intellectual life in this period, and it was difficult to find a neutral meeting ground where conservatives, radicals, liberals, and others could exchange views. For this reason, the Literary-Scientific Society became a success, soon enrolling some three hundred members, scheduling dozens of lectures, discussions, readings of original works of criticism and poetry, etc. The Society also eventually built a library of some five thousand volumes under the direction of Sergei Oldenburg, which included controversial books freely accessible to members and often more up-to-date than the university library. The Society was one of the few places in the university where students could display initiative in a relatively open atmosphere. Most university classes, especially in the first few years, were large lecture courses in which students tended to participate passively. In the Literary-Scientific Society students could gain confidence as public speakers, presenting their own original work as well as discussing important ideas in European and Russian intellectual life.

In the work for the Literary-Scientific Society, Vernadsky developed the organizational talents for which he and other members of the Oldenburg circle became known in later years. As Ivan Grevs expressed it, they "possessed a rare mastery in attracting people to an organizational task. They displayed here their energy and their ability to assess the value of a particular candidate and to direct his talents to that task most useful to the society."[105] The ability to attract people and judge their talents accurately was to prove invaluable in Vernadsky's later career as a scientific administrator and founder of a series of important institutes within the Academy of Sciences, just as it was an important experience for Sergei Oldenburg, who later became the Permanent Secretary of the Academy of Sciences and was one of Vernadsky's closest friends and associates in future decades.

Beginning in 1883–84, Vernadsky also found an extracurricular outlet in the St. Petersburg Committee of Literacy. Prince Shakhovskoi interested many members of the Oldenburg group in this activity. Their aim was to prepare reading materials "for the people." There is no evidence that Vernadsky was directly involved in teaching workers to read and write. Rather, the group he joined drew up reading programs for adult education, helped to establish popular lending libraries and was concerned with preparing and disseminating modern knowledge and works discussing moral issues at a level accessible to the newly literate who had little or no previous formal education. Such activities, in an era of "small deeds," were very widespread among Tolstoian and populist circles at the time and eventually burgeoned into a large industry which published many works on popular science, history, tracts on moral issues, public hygiene, diet, modern farming methods, and sanitation.[106] These books and pamphlets were published in cheap editions and circulated among working class and peasant elements of the population. One of Vernadsky's classmates at the university in these years, Nikolai Rubakin, a political radical of populist sympathies, later made a career of writing such works of popular science, becoming a kind of Isaac Asimov of Russian popular culture who eventually saw millions of his books in print. Rubakin founded one of the largest circulating libraries in St. Petersburg and became a cultural institution in his own right, corresponding with thousands of workers and others, advising them on their reading and programs of self-education.[107]

Vernadsky himself was not suited by temperament or personality to become primarily a popularizer or cultural worker of the same stamp as Rubakin. However, judging by one of his diary entries for 1884, the future scientist attached great importance to such work as a means of making widespread a scientific outlook on nature. Such an outlook was for him an indispensable part of any political transformation of nature and society. "The more deeply scientific information penetrated the masses, the better," Vernadsky wrote in his diary on May 21, 1884. "Our common idea must be that the people must understand their own strength and rights, it must be to lead the people to a consciousness of the need to govern themselves. . . . only when the majority of the people understand their own situation and their own strengths, only then will a more intelligent overthrow of the presently parasitical government be possible."[108] Practical knowledge and a scientific world view, he felt, went hand in hand with the establishment of a popular government which would "exist for the benefit of the people and not where the people exist for the government." Vernadsky felt that the task was a gradual one that required great patience and effort: "Up till now the people have not been touched by scientific knowledge. Old ideas and an old world view, many centuries ago outmoded by science, still possesses them. Very, very slowly, with a great effort, scientific knowledge becomes part of the popular consciousness."[109]

Although he did not use the expression, Vernadsky clearly thought in terms of a peaceful cultural revolution, carried out by the educated members of society, that would lead to a political transformation, a more democratic system, rather than the reverse order of events, a political revolution followed by cultural transformation, which many political radicals believed necessary.

"What is the idea that links us all together?" Vernadsky asked in his diary and answered: "To try to spread a scientific worldview among the people."[110]

How much time Vernadsky actually devoted to these extracurricular activities is unclear. It was possible to make a full time career of them, as many of the Oldenburg circle were prone to do, at the expense of their formal studies. We know that this was not the case with Vernadsky, who was dedicated to becoming a professional scientist of high calibre. But even if his thoughts here expressed more words than concrete deeds, he had set himself a lifetime program and remained consistent to these beliefs through the remainder of his life.

Participation in the Committee on Literacy had a fringe benefit he did not enjoy in his formal and informal university associations. The Committee attracted to its work many young women. Women were excluded from attending the university, but through the Committee several members of the Oldenburg circle met their future wives, including Vernadsky.[111] Nataliia Egorovna Staritskaia was three years Vernadsky's senior, a serious, rather plain woman, judging by her photographs from the period.[112] According to Vernadsky, she bore a striking resemblance to a painting in his father's study of his first wife, Maria Shigaeva. There was perhaps a certain amount of idealization involved in this resemblance, since as we know, Vernadsky at this time was critical of his own mother's lack of intellectual interests and moral values shared with his father. No doubt he felt a great interest in a marriage that by contrast would be more companionate, as he believed his father's first marriage had been. At any rate, he enjoyed Natasha's company for some time before proposing, escorting her home from meetings of the literacy group and sharing many common interests in literature and European thought. Natasha was fluent in French and German, reading widely in the literature of those countries as well as her own, and she later translated a number of Vernadsky's scientific works into these two languages, giving his ideas a much wider audience.

Natasha came from a cultured family of university background; her father was a high official in the Tsarist government known for his rare honesty and integrity.[113] Egor Pavlovich Staritsky was the head of a legal department in the State Council. But he was also an internal critic of the regime who had helped to carry out the legal reforms of the 1860s and was devoted to making Russia a *Rechtstaat,* a government in which the rule of law outweighed the personal whim of the autocrat. He was very much opposed, therefore, to the retreat from law which he saw manifested in the reaction of the 1870s and the counter-reforms during the reign of Alexander III after 1881. Vladimir recognized a soulmate not only in Natasha, but in her father as well, something he had missed since his own father's death. He took to spending a good deal of time at their apartment in St. Petersburg and their dacha in the Finnish countryside at Terioki. However, when he finally proposed to Natasha, she at first turned him down, protesting that the difference in their ages was too great.[114] Her resistance did not last long, however, and in the fall of 1886, they were married, an affectionate relationship that lasted—with no evidence of serious lapses—for fifty-six years.

The wedding ceremony proved to be a bone of contention between the young couple and Vernadsky's mother, however. She insisted on a lavish and expensive ceremony, with engraved invitations, frock coats and wedding gowns, hired carriages and an orchestra. Vladimir and Natasha finally gave in, although such lavishness contradicted their principles and those of the Oldenburg circle. When the other members of the Oldenburg circle received invitations, however, they heartily disapproved of the conspicuous expense involved and boycotted the ceremony.[115] The moral censorship of this group could be very sharp, as another member discovered when he fathered an illegitimate daughter and abandoned her mother, who placed the child in an orphanage. The other members of the Oldenburg group were so strong in their condemnation of his actions that he changed his mind and agreed to raise the child himself.[116]

The young married couple were soon back on speaking terms with their friends, who became frequent visitors at the modest three room apartment the Vernadskys rented in a lower class section of St. Petersburg. One of the Oldenburg group was even delighted when he discovered that the Vernadskys, quite innocent of the fact, had taken an apartment next to a notorious legal brothel. Their student friends jokingly approved of the choice, since they felt that the police would not think to put their activities under surveillance in such a disreputable neighborhood.[117] Apparently the police expected "subversives" to live in better neighborhoods. The need to avoid police surveillance seemed especially important at the time, since in 1886 Vladimir had become president of the United Council of Student Organizations (zemliachestva).

The zemliachestva were a series of organizations which were formed by students who came from the same area in the provinces. The organizations were considered illegal by the government. Somewhat larger than most student circles, the zemliachestva attempted to provide moral and material support to their members, many of whom arrived in St. Petersburg with no contacts among the local population and therefore tended to look to fellow students from back home for help and friendship.[118] Although Vernadsky was from St. Petersburg, his identification with the provinces and his association with the Ukraine put him on friendly terms with the provincials. Being elected President of the United Council of these organizations was a tribute not only to his leadership ability, but also to the trust he enjoyed with a broad spectrum of student opinion.

The year following Vladimir's marriage proved to be far from quiet. In fact, it marked a high point of repression against student activists like Vernadsky and others. The axe fell first on the Literary-Scientific Society. This society might have continued to exist for some years, except for the events of March 1, 1887. On that day, the anniversary of the assassination of Alexander II, a group of students were arrested in a plot to kill Alexander III. Some students were apprehended while standing on the street waiting for the Tsar's carriage to pass, bombs and revolvers in hand. Others were arrested later in their living quarters. During the investigation that followed, it turned out that many of the student conspirators were members of the Literary-Scientific Society, including Lenin's brother, Aleksandr Ulianov. Ulianov had, for a time, been secretary of the Scientific section of the society, but had resigned several months before the assassination attempt.

Vernadsky and others members of the Oldenburg circle had known Ulianov, who had entered the university as a zoology student in 1883 and therefore was still an undergraduate in 1887. Ulianov had attended meetings in their apartments but was known as a secretive person, who revealed little of himself or his feelings. Ulianov was one of a number of radicals who attended the activities of the Literary-Scientific Society, but the latter organization was not directly involved in the plot. Vernadsky had some hint of Ulianov's radical views from occasional remarks but was not aware of the extent to which he and his friends would act on their opinions. Before his arrest, Lenin's brother, through an intermediary, had left a trunk in Vernadsky's keeping at the Mineralogical Museum, where Vernadsky served as curator. Soon after the assassination attempt, Vernadsky checked the trunk and discovered that it contained dynamite fuses. Vernadsky, who saw violence as counterproductive and considered himself a pacifist like Tolstoi, quietly removed the trunk one night shortly after Ulianov's arrest. With his friend, Sergei Oldenburg, Vernadsky rowed out into the middle of the Neva River and sank the incriminating object.[119] During the police interrogations, Ulianov claimed chief responsibility for plotting the assassination. Although thirty-six students were arrested in the affair, only Ulianov and four of his associates were executed.[120] However, serious consequences for the university and for Vernadsky personally were not long in coming.

Discovery of the assassination plan provided a convenient pretext to abolish the Literary-Scientific Society, despite its merely incidental connection with some of the plotters. A few months later another pretext was found to fire its faculty sponsor, Orest Miller. At the end of the academic year, Miller was permanently removed as a professor at the university and died in official disgrace two years later. As the minister of education put it at the time, "it is better to have a professor of mediocre capabilities than an especially gifted professor who, however, exercises a deleterious influence on the minds of the students."[121] The irony in this is that Miller had been devoted to keeping radical political activities out of the Society, probably an impossible task under the circumstances, however, since any free discussion of ideas had political overtones in the repressive atmosphere.

A Slavophile in his personal beliefs and an ardent student of contemporary Russian literature, a particular admirer of the conservative novelist Dostoyevsky, Orest Miller was known to revere the memory of the assassinated Tsar Alexander II and was a devoted Russian Orthodox believer. However, his religion was not one of passive resignation. Like the Slavophiles, he believed in the right of the Russian people to enjoy an independent cultural and social life and to express freely their opinions to the Tsar. He also believed in acting on his Christian faith through good works and acts of mercy. Among other activities he had organized a dining room and cooperative loan association for needy students before becoming faculty sponsor of the Literary-Scientific Society. As a result, he and the society were very popular among the students, including radicals—indeed, too popular for the comfort of officialdom.

A short, balding man who lived in a poorer section of St. Petersburg, Miller was known more as an enthusiastic teacher than as a brilliant scholar. He taught not only at the university, from which women were barred in this period, but also at an unofficial university which feminists and their friends had organized in the capital. Miller had a reputation for being accessible to students in trouble, lending them both his money and his good advice when asked. Many of his more formal colleagues, not to mention the Ministry of Education, considered such familiarity with students inappropriate to his position.[122] However, for graduate students and budding young academics like Vernadsky and his friends, Orest Miller was a model of the good teacher. The memory of his helpfulness and enthusiasm for learning remained strong in a university where many of the senior faculty were known for the low quality of their teaching and their inaccessibility to students. Miller's example probably also strengthened Vernadsky's interest in religion and belief in cooperating with religious people who worked for progressive social change. Although excerpts from his student diary indicate that Vernadsky did not consider himself an Orthodox Christian believer, the scientist retained a lifelong interest in religion, which associations with intellectuals like Orest Miller (and later the mystics Vladimir Solovev and Sergei Trubetskoi, one of his closest friends at Moscow University in the early 1890s) only strengthened. Indeed, some of his fellow students at the university thought that Vladimir had a "leaning toward spiritualism" in these years,[123] and there is evidence that he did at times, that there were both rationalist and mystical sides to his nature, although rationalism usually predominated. The nature of his philosophical outlook in this respect is not altogether clear, but one fact is certain: Vernadsky, unlike most Russian radicals of the time, had great respect for the religious side of mankind's nature (even as a stimulus to science) and worked well with religious figures who shared his own ideals for a more progressive and democratic Russia.

The events of March 1, 1887, gave additional impetus to those elements in the Tsarist government who wanted a general purge of university faculty and students whose loyalty to the autocracy was suspect. Not only were some professors dismissed, like Orest Miller, others, like the famous embryologist Mechnikov, left Russia for good. Mechnikov took up a permanent position at the Pasteur Institute in Paris.[124] Reactionary elements in the government thought that student unrest in the universities arose from the increasingly "democratic" social composition of the student body. (The proportion of noblemen in the universities fell from approximately 65 percent in 1855 to 45.8

percent in 1875 and continued to fall in the 1880s.) "Therefore, the Ministry of Education was given the task of reforming the student body so that it would consist of noblemen, be homogenous in level of wealth, and constitute a closed caste of future officials monolithic in its loyal ideology."[125] The new rector of the St. Petersburg University was M. I. Vladislavlev, a conservative drawn from the History-Philology chair in philosophy. One of his chief claims to academic distinction was a theory which claimed that a man's positive psychological qualities varied directly in proportion to his wealth. That is, the greater one's wealth, the higher one's positive qualities.[126] This theory fit nicely with the official policy of the minister of education, Delianov, who in 1887 issued his famous decree raising tuition to the universities in order to discourage "the children of cooks, coachmen," and other lower class elements from attendance.[127]

That summer of 1887, Vernadsky sent Natasha to spend the summer with her parents in the Finnish countryside, since she was pregnant at the time, while he took part in a geological expedition. Before long, however, he was summarily recalled to St. Petersburg by the rector of the university. When Vernadsky appeared before the rector, Vladislavlev was examining a dossier which contained a denunciation of Vernadsky by an anonymous informer. The latter reported hearing a conversation between Vernadsky and his friend Krasnov in the university's Mineralogical Museum during which Vernadsky supposedly expressed approval of the terrorist attempt on the Tsar's life. At the time the conversation reportedly took place, Vernadsky pointed out, Krasnov was already in Western Europe pursuing his graduate studies in plant geography. "Besides, Your Excellency," Vernadsky later recalled saying, "you may not know that I have inherited five hundred desiatins of land [about 1450 acres] from my father. Psychologically, therefore, I would be unable to do anything which would contradict those qualities that belong to me by virtue of my material position."[128] Such an inheritance (which Vernadsky owned until dispossessed by the revolution of 1917) placed him among the top twenty-five percent of the hereditary nobility by wealth (although, in fact, the estate of Vernadovka in Tambov province never produced more than a modest income).[129] At any rate, 1450 acres of black earth seemed an impressive landed estate to Vladislavlev, who found himself hoisted by the petard of his own "psychological theory." The rector had to agree that Vernadsky should be more careful about possible informers among his acquaintances. However, this was not the end of the matter.[130] It was not long before the minister of education himself, Delianov, requested Vernadsky's presence. The minister was more curt, and left no time for questions or explanations. He informed Vernadsky that his presence at St. Petersburg University was no longer desirable, for reasons which he did not care to discuss, and he summarily asked for Vernadsky's resignation "for family reasons." In this way, no blackmark would appear on his civil service record.[131]

Vernadsky did not resign immediately but rode out to his inlaws' summer cottage at Terioki to report Delianov's request. His father-in-law, Egor

Pavlovich, was irate. Natasha agreed with her father that he should intervene to stop this arbitrary action. Early the next morning, therefore, Staritsky left for the capital and appeared at Delianov's office, demanding an interview. Impressed by a calling card that informed the minister of education that Egor Pavlovich Staritsky, president of the Department of Laws of the State Council, wished an audience, Delianov immediately welcomed the elderly official to his office. Staritsky, decked out in his Imperial uniform, with gold braid and the insignia of high bureaucratic rank, demanded an official explanation why his son-in-law should be forced to resign: "There is no law, your Excellency, if I recall, no such statute that permits the firing of a civil servant without explanation of the reasons." (Vernadsky held civil service rank as a graduate student and curator of the Mineralogical Collection, and therefore deserved treatment under the rules of proper civil service procedure.) This blatant violation of proper legal procedure touched a very sensitive spot in Staritsky's nature. Delianov replied that he was acting at the personal command of the emperor "to clean out the universities of all unreliable elements."[132] He then mentioned Vernadsky's activity in illegal student organizations, going back to his undergraduate days. But what was most suspicious, he felt, was Vernadsky's refusal two years earlier, of the opportunity for graduate study in Western Europe. Why did he wish to stay near the university? The implication was that he wished to remain in St. Petersburg, since he was involved in subversive activities. Staritsky explained that following the death of his father, Vernadsky had been asked by his mother to remain in St. Petersburg to help with family matters and provide some consolation for her, since he was their only son and now head of the family. Delianov seemed satisfied with this explanation and agreed to Vernadsky's remaining in the university, provided that he accept a new invitation to go abroad for graduate study in Western Europe. Since Natasha was pregnant, the trip could be postponed until the spring of 1888, but no later.[133]

On September 1, 1887, Natasha gave birth to a son, the future historian of Russia George Vernadsky, named in honor of his maternal grandfather.[134] Thanks to the personal intervention of his father-in-law, Vladimir Vernadsky's future scientific career in Russia had been rescued and kinship ties with his wife's family strengthened. In the spring of 1888, a new chapter began with his departure for the capitals of Western science.

CHAPTER

2

SCIENCE AND SOCIETY
The Origins of Vernadsky's Scientific
School, 1888–1905

Although by the end of the nineteenth century most Russian professors received their degrees from a Russian university, it was still traditional to go abroad for a year or two of study in a particular specialty sometime between receiving the undergraduate diploma and finishing an advanced degree at a Russian university. Initially, Vernadsky chose the University of Naples, drawn back to Italy in part by the memory of travels there with his father, in part by the reputation of Professor Scacchi of that university, a well known crystallographer and mineralogist. However, this initial choice proved to be a poor one, an indication of inept advising by Vernadsky's professors at St. Petersburg University and by the Imperial Russian Ministry of Education, which had to approve his study plans in Western Europe.

The truth of the matter was that the fields of mineralogy and crystallography, in which Vernadsky had chosen to prepare himself, had virtually disappeared as well-developed disciplines in Russian universities. Mineralogy, for example, was taught primarily by professors whose primary interest was in some other field, such as the soil scientist Dokuchaev at St. Petersburg University or the geologist, A. P. Pavlov, who taught the subject reluctantly at Moscow University. The last real university mineralogist had died in 1887, a rather startling fact when one considers the mineral wealth of the Russian Empire, and the need for much greater knowledge of the country's resources before that wealth could be developed.[1] At any rate, Vernadsky and his wife left St. Petersburg in March 1888 (his baby son remaining with his in-laws at their dacha in Finland) and set out for the University of Naples, where he planned first to learn methods of analyzing crystals with Scacchi, who had a world-wide reputation in that field. However, Vernadsky soon discovered that Scacchi was senile and that active research in crystallography was no longer being conducted at this Italian university.[2]

Within a few days the Vernadskys had left Italy for Germany where Vladimir decided to study with the foremost German crystallographer of the time, Paul

Groth, Professor of Mineralogy at the University of Munich and Director of the Bavarian State Mineralogical Collection. Groth was a formidable presence in his field, editor of the *Zeitschrift für Kristallographie und Mineralogie,* which he founded in 1877, and author of the principal German text on physical crystallography, which went through a number of editions between 1876 and Groth's death in 1927. Groth's "most important contribution to science" was "his explanation of the connections between chemical composition and crystal structure."[3] He was particularly interested in the paragenesis of minerals, an area also of special concern to Vernadsky and no doubt one of the major reasons the young Russian decided to come to Munich. The other reason doubtless was Groth's inventiveness in developing scientific instruments for the study of crystals. "In 1871 Groth had improved the polariscope, the stauroscope, the axial-angle instrument, and the goniometer and combined them into a universal instrument."[4] With the help of such advanced instrumentation, which German industry was able to reproduce rapidly and make available on a wide basis to the scientific community, Groth and his students in Munich systematically began to investigate "the optical, thermal, elastic, magnetic, and electrical properties of crystals."[5]

In the spring of 1888 Vernadsky set about to learn the use of such scientific instruments, working on a topic assigned to him by Groth, together with one of Groth's German graduate students, a man by the name of Mutman, who Vernadsky complained did little actual work on the project.[6] In addition to working with Groth, Vernadsky took advantage of the presence in Munich of Professor Zonke who made available to graduate students his well equipped physics lab. Zonke was working on the theory of crystallization, and therefore work with him was especially valued by Vernadsky.[7] The application of methods of physical and chemical analysis to minerals was the key to greater knowledge in mineralogy, and Vernadsky felt inadequately prepared by the lack of modern equipment and methods in his education at St. Petersburg University.

At the end of the spring semester in 1888, Vernadsky wrote a lengthy letter to his mentor Dokuchaev in St. Petersburg, reporting on the progress of his studies: ". . . in general I am very satisfied with my work here. At the outset I devoted all my time to becoming familiar with research methods—optical and crystallographic; I made measurements of triangles of different crystals of various sizes with various instruments . . . After two or three weeks Groth gave me a small independent project which I am conducting together with his research assistant, Mutman: determining the optical anomalies of one complex organic compound. . . . However, I am not attempting to arrive at any conclusions about optical anomalies, but undertook this work because it provides an opportunity to learn regular methods of research. . . . I will not finish the project this semester but will continue it next semester."[8]

Vernadsky went on to explain that during the following semester he would also be studying the capillarity of crystals. "Groth completely approves of this . . . he has talked it over with Zonke, and from October on, I will work in

Zonke's physics lab, at first doing practice work with capillary phenomena." Vernadsky planned to test the capillarity of crystals not only with liquids but with gases, but he expressed doubt about what results might be obtained. His letter apparently aroused some concern on Dokuchaev's part. The latter perhaps feared that his pupil's well intentioned curiosity about a wide variety of natural phenomena was leading to a lack of direction. At any rate, in his reply on June 3, Dokuchaev wrote briefly and to the point, inquiring: "Write please, more precisely in *what specifically* will be your *special* task—that is to say your personal theme?"[9] Vernadsky had left Russia having completed his exams for the first graduate degree *(magister)* but without having defined a clear topic for his thesis. This letter from his major professor marked the beginning of a steady drumbeat in the correspondence between pupil and mentor in which Dokuchaev continually reminded his twenty-five year old graduate student that completion of his master's thesis was essential for Vernadsky to obtain a decent university post in Russia.

Dokuchaev's pressure to complete the professional degrees was indicative of what was to become a major source of tension in Vernadsky's life. Some might have viewed Vernadsky's broad interests as a tendency toward dillettantism but many members of the nineteenth-century Russian intelligentsia obviously considered such broadness to be a positive value, a desire to know not only a variety of scientific fields and techniques but to travel widely, remain *au courant* with trends in art and music, read widely in literature, philosophy, and history and combine all of this with a life of service to the people of the Russian Empire. Vernadsky's voluminous correspondence is indicative of the creative tension with which he lived, leading at times to intense headaches lasting two weeks and longer. Anxieties over his career and family life (particularly the health of his wife, mother, and father-in-law), as well as social pressures from friends of his undergraduate days to work harder in the zemstvo movement and other activities to raise the cultural level of Russia, might have broken a person of less sturdy constitution. Vernadsky emerges from these years as a man of great internal strength, but he paid a price not only in terms of scientific productivity, but happiness as well. Not only a sense of commitment unusual in its conscientiousness, but also a very strong marriage, in which only occasional cracks and strains show through in the many letters to and from his wife in these years, helped to carry him through.

Vernadsky's wife returned from Munich to her parents' dacha in Finland during June of 1888, to be with their infant son. In his letters to her over the next six months, we catch a glimpse of Vernadsky's scientific and personal development. The tone is at times that of a youngster in a candy shop, with so many delicacies to sample he scarcely knows where to begin. Despite his dislike of a "certain German scientific chauvinism,"[10] Vernadsky quickly made a number of friends in the scholarly community of Munich. For example, while attending Groth's lectures he met a group of students in the biological sciences, whom he admired for their idealism and enthusiastic pursuit not only of the details of scientific knowledge but of larger philosophical questions as well. Chief among this group was Hans Driesch, a graduate student of Ernst Haeckel, the principle defender and promoter of Darwinism in nineteenth-century Germany.

Later in the 1890s, Driesch was to become the center of one of the most important controversies in nineteenth century biology, the controversy between the vitalists, who argued that there was a non-definable "life force" that directed the development of organisms, and the mechanists, who argued that material causes can be found at the basis of all organic and inorganic phenomena in nature. Vernadsky drank "bruderschaft" with Driesch, and there is evidence that the Russian scientist kept up a correspondence with the young German idealist during the 1890s, although copies of these letters have not come to light.[11] When Driesch began to publish monographs in the 1890s arguing for vitalism, there is also evidence that Vernadsky read at least one of these works and was skeptical about the argument. All of this is significant for several reasons: it is further evidence of Vernadsky's fascination with broader philosophical questions involving science; and secondly, Vernadsky's attitude toward vitalism was later to become an issue in his polemics with Marxist-Leninist philosophers in the Soviet Union during the 1920s and 1930s who at times accused Vernadsky himelf of being a vitalist. In notes for an autobiography he wrote toward the end of his life, Vernadsky remembered his friendship with Driesch warmly, but stated that he had always viewed vitalism with caution. No other evidence has come to light that would tend to contradict his later memory, although we know from Vernadsky's personal diaries that he was much attracted to an idealist philosophical position in the early 1890s.[12]

Although Vernadsky met several of his German friends, including Driesch, as the result of attending lectures at the University of Munich, in general he became very critical of the lecturing style of German professors and began to attend less and less, spending more time in the libraries and particularly the laboratories of Munich. By November he was writing Dokuchaev: "I am attending few lectures; in general the lectures of German professors are so elementary they cannot compare with ours."[13] Despite this note of criticism, it is also clear that Vernadsky experienced great intellectual growth during these months. He spent two months during the summer of 1888 traveling through Germany, Austria, Switzerland, and France, in part with another German friend, Zeitel, who taught mineralogy at the University of Munich. They toured the libraries and mineralogical collections of Europe and visited some of the principal mineralogical deposits of these countries. Such scientific field trips had not been a part of his education at St. Petersburg University, just as the laboratory work and quality of instrumentation there had been inadequate, Vernadsky now realized.

For seven years, Vernadsky complained to his wife, he had been engaged in scientific research and was only now learning proper methods of observation and experimentation. "At the university [in St. Petersburg] I threw myself with passion into the natural sciences, but I read more than I observed; I have really only begun to observe things properly during the last year."[14] Such practical mineralogy was one of the most valuable things he was learning in Western Europe. Observation and laboratory research at times led him into a mood of great enthusiasm when writing his wife, at other times into depression or boredom. The research he continued with Mutman in the fall of 1888 involved a vast number of precise measurements for determining the capillarity of various crystals. Vernadsky was apparently a very conscientious laboratory researcher who continually checked and rechecked results, always worrying that perhaps he had made some mistake in detail that would change the entire result.[15] What started out in the spring as a practical exercise to learn certain research methods, by fall began to turn up unexpected results which attracted the enthusiastic attention of Groth and Zonke.

Vernadsky was scheduled to leave Munich by January or February of 1889, to study in Paris with Fouqué and Le Chatelier. Groth was obviously impressed by his Russian visitor and tried very hard to convince Vernadsky that he need spend no more than a month in Paris to learn all that the French could teach him, whereupon he should return to Munich to work more closely with Groth. Groth also asked Vernadsky to write reviews for the *Zeitschrift* which he edited, another indication of his confidence in the young Russian's competence. Both Mutman and Groth urged Vernadsky to publish the results of his laboratory findings in Munich, but Vernadsky was reluctant to do so, apparently viewing his results with caution. He was hesitant to risk drawing too general a theoretical conclusion from work he was not sure he had yet mastered.[16] He also resisted his mentor Dokuchaev's urgings that he use this research for his dissertation.

Besides, the capillarity of crystals was not the only subject he wished to investigate while in Western Europe, and he was not ready yet to settle down to a single specialized topic. He viewed his time in the West as an opportunity to delve into a number of topics of interest, work with a variety of distinguished scientists, learn as many new research techniques as possible, and become acquainted with the scientific and general cultures of several countries. Besides, as we know from his earlier years, Vernadsky lost no love on the Germans, even while he was able to learn a great deal from them. He was more a Francophile and wanted to live and study in Paris for a time. Vienna was also on his list of destinations, but he finally had to give up plans to study there for lack of time, spending the remainder of his stay abroad in Paris, from February of 1889 until his return to Russia in the spring of 1890.

Toward the end of his stay in Germany, it is clear from his letters to his wife, Vernadsky was bored with the work he was doing on the capillarity of crystals. Natasha had apparently gotten the impression from many of his earlier letters, which dwelt at length on various hypotheses which Vernadsky wished to test, that science was a fascinating, all-absorbing enterprise, involving ever new and entertaining ideas. Vernadsky frequently apologized to his wife at the end of a long letter in which he spent most of the time discussing his scientific work that he was surely boring her, since from her own letters it is clear she was far more interested in family matters and in people than she was in the details of science. In fact, one of the few sharp disagreements in their correspondence was over the lack of detail on everyday life and acquaintanceships in his letters, for which his wife scolded him and he defended himself during the winter of 1890.[17]

On February 19, 1889, just before leaving for Paris, Vernadsky wrote his wife: "You are mistaken in your opinion about the interest of scientific work; what is interesting are generalizations, perhaps a new interpretation of results. It is interesting to read [about all of this]. The interest in scientific activity consists in the investigation and clear understanding of goals, but scientific activity itself is not easy and most of the time is devoted to mechanical, boring work."[18]

In the eyes of his teachers, both in Russia and Western Europe, Vernadsky was a highly competent, trusted laboratory and field researcher. Groth not only wanted to keep him longer in Munich, in 1888 Dokuchaev also named Vernadsky to the newly formed Soil Commission of the Free Economic Society which was charged with carrying out an extensive field research project on the soils of Poltava province in the Ukraine, an indication of the high regard in which that eminent soil scientist held his pupil. But clearly what interested Vernadsky most about science was the broad picture, the picture it could give of the laws of nature, the formation and development of natural phenomena. For example, while traveling in Switzerland during the summer of 1888 with the German mineralogist Zeitel, Vernadsky had one of those experiences that were to punctuate his life, giving direction to his work, events that Vernadsky experienced with great intensity.

While viewing the stars through the clear night air of the Alps, he more clearly defined to himself his goal as a mineralogist: "To collect facts for their own sake, as many now gather facts, without a program, without a question to answer or a purpose is not interesting. However, there is a task which someday the human mind will solve, one which is extremely interesting. Minerals are remains of those chemical reactions which took place at various points on earth; these reactions take place according to laws which are known to us, but which, we are allowed to think, are closely tied to general changes which the earth has undergone as a planet. The task is to connect the various phases of changes undergone by the earth with the general laws of celestial mechanics. I believe there is hidden here still more [to discover] when one considers the complexity of chemical elements and the regularity of their occurrence in groups. . . ."[19]

This concern for the earth's place in the cosmos and the desire to explain the laws of the earth's development as a planet within the context of laws of cosmic development was to become a persistent theme in Vernadsky's later scientific career. What he lacked in steady devotion to a single scientific specialty, he made up in his ability to ask important questions and in his interest in general theories of development. While he may have been a competent experimentalist, his real interest was in the development of scientific theory; and once he had begun to probe a particular problem or field, his tendency was eventually to leave the drudgery of experimentation and the collection of facts to others, while he transferred his interests to some new scientific frontier. This is not to deny that Vernadsky was capable of doing very good experimental and field work, collecting new facts and organizing them into a coherent explanation. It is simply to say that he was primarily a theorist and a generalist, who left narrow specialization to others, especially to his own pupils and associates, who came to form the Vernadsky school in Russian science. As in later life, Vernadsky's early career was marked by a tendency to jump from theme to theme and to alternate from detailed experimental and observational studies to broad theorizing and speculation. Vernadsky himself feared that at times more down-to-earth scientists such as Groth "take me for a fantasizer."[20]

In February 1889, Vernadsky moved from Munich to Paris. He arrived in Paris as the French capital was preparing to celebrate the centenary of the French revolution. Gustave Eiffel's tower was the sensation of the moment as the French prepared to open an international exposition. Vernadsky's mentor Dokuchaev was invited by the French to prepare an exhibit of his work in soil science. He wrote Vernadsky asking if the latter would act as his personal agent, supervising the construction and mounting of the exhibit, a task Vernadsky undertook with energy and particular pride when Dokuchaev was awarded a gold medal for the exhibit. Or as Vernadsky himself expressed it dryly in a letter to his mentor, Dokuchaev had won the right to "purchase a gold medal" from the French.[21] Vernadsky in return for his official functions at the exposition received a free pass to enter any of the attractions and was able to spend a great deal of time examining the variety of scientific and technical exhibits from many nations. Paris in 1889 held many such distractions, and Vernadsky concluded that he would have to postpone writing his dissertation until after his return to Russia, as he informed Dokuchaev in February.[22] Nonetheless, it was in Paris that he finally found a dissertation topic about which he felt both enthusiastic and confident that it would yield original results. This happened while working with Professor Le Chatelier of the Mining Academy and Professor Fouqué of the Collège de France.

Henri Louis Le Chatelier was the model of a French technocrat. His father had been inspector general of French mines and had overseen the construction of much of the French railway system. Henri Louis followed in his father's footsteps and provided for Vernadsky a role model of how a scientist could aid in the development of his country by working first as a mining engineer and then as a professor of industrial and general chemistry. Upon receiving the doctorate in physical and chemical science in 1887, he eventually held the same post as his father, inspector general of mines, and became a member of the French Académie des Sciences. "During his life he held increasingly important positions on a large number of commissions and boards to advise the government on scientific and technical questions. Among these were the Commission on Explosives, the National Science Bureau, the Commission on Weights and

Measures, the Commission on Standardization of Metal Products, the Commission on Inventions," etc.[23] He founded and edited the *Revue de métallurgie* and trained scores of graduate students. When Vernadsky met him he had just reached the age of forty and was establishing himself as one of France's leading scientists.

Le Chatelier was a brilliant teacher and spent a great deal of time conversing with his students. Vernadsky remembered long talks with him on a variety of topics and considered him one of the most brilliant minds he had met in Western Europe. The French scientist had certain pet enthusiasms which influenced Vernadsky's later work. For example, Le Chatelier was a great believer in the application of mathematics to chemistry and other areas of science, and it was from him that Vernadsky first heard about the work of the American mathematician J. W. Gibbs, which he was later to apply to some of his work in mineralogy and crystallography.[24] While investigating conditions of equilibrium in reversible systems, Le Chatelier discovered the work of Gibbs. He "recognized that the highly abstract mathematical form of Gibbs' presentation had prevented most chemists from understanding its chemical applications, and he did his best to spread knowledge of these in France. In spite of his efforts chemists did not accept it quickly, but he himself utilized it in the rest of his work."[25]

Vernadsky, although he always felt his inadequacy in mathematics, also became an enthusiast for the mathematization of science, at first in mineralogy and crystallography, and later in geochemistry and biology after 1917. Although his work with Le Chatelier was only one of the influences in that direction, it was clearly an important one, in Vernadsky's own view. The French scientist also influenced Vernadsky in another respect. He advised caution in Vernadsky's tendency to take too broad a theoretical and even speculative approach in science. Le Chatelier was a critic of scientific abstractions and tried to use only concepts that were thoroughly grounded in empirical observation and particularly in experimentation. He even went so far as to eliminate the use of such abstractions as "force" and "atomic theory" in his teaching. While Vernadsky remained far more theoretically oriented and willing to use abstract concepts during his career than Le Chatelier, the French scientist's influence can be felt in Vernadsky's later preference for the use of concepts in science which he called "empirical generalizations," based soundly on experimental and field work.[26]

However, the most immediate influence Le Chatelier had on the young Russian scientist was in helping Vernadsky find a dissertation topic. Le Chatelier was interested in phenomena of polymorphism, that is, the ability of some chemical compounds to appear in several different crystalline forms. He was also an expert on the most common minerals found on earth, the silicates. It was in this area that Vernadsky found his topic, while working with Le Chatelier and his other advisor, a less brilliant or at least less talkative man, Ferdinand André Fouqué, Professor of Natural History at the Collège de France. Fouqué was a specialist on the creation of synthetic minerals in the laboratory, and it was in the process of trying to synthesize sillimanite that Vernadsky arrived at a hypotheses about the structure of silicates. He was able to combine this with the work he was doing with Le Chatelier to form the basis for his dissertation.

As he wrote Dokuchaev: "My basic idea is the proposition that silicates containing aluminum, iron, oxide, chrome oxide, and boric anhydride are not salts of some kind of siliceous acids but salts of complex acids—silicic-alumina, silicic-boric, etc. Even if I do not succeed in obtaining complete proof it seems to me that just the posing of the question in this form can help unravel this and other problems connected with silicates. . . ."[27]

Vernadsky found French scientific organizations to be smaller and less well equipped than those in Germany, but felt that they made up in friendliness and hospitality for what they lacked in other respects. "I have begun work here with Fouqué in his laboratory at the Collège de France on the laboratory synthesis of minerals. . . . All such work in Paris is scattered about among a whole series of laboratories; part of my experiments I must do with Fouqué, others only with Friedel at the School of Mines; finally, both Fremy (at the Jardin des Plantes) and Tomepei (at the Sorbonne) have their own labs equipped for this type of work. The technology for this work is not sophisticated. . . . I have already done several preparations . . . and perhaps can attempt special work on a group of little studied compounds, alumina with silicon dioxide. I hope that through synthesizing and studying the conditions under which these compounds are formed to be able to explain their chemical composition (the group of andalusites, disthene, sillimanites, dumorterites, etc.). If I cannot carry out such independent work, then I will have to end my visit, since I will soon finish learning the actual methodology itself."[28]

As it turned out the work went well and by April, even with all his other distractions, Vernadsky was reporting to his mentor, "I have begun work with Fouqué on the compounds Al_2O_2 with SiO_2 and this has evidently produced results; it seems that I have produced sillimanite, until now never before produced synthetically."[29] By October, Vernadsky was thinking of extending his stay in Paris until the spring of 1890, partly because his wife's health was improving there (she had arrived with their infant son some months before) and partly because his work was progressing so well. He felt he needed the time to complete the experimental work for his dissertation while he had the momentum. "I will work on my dissertation. I have obtained rather unexpected results and I am thinking about the problem of dimorphism in general." What Vernadsky had produced in the laboratory were synthetic sillimanites in two different crystalline forms. If his results checked out, he told Dokuchaev, "this will be confirmation of dimorphism of the entire group of olivines, that is the group of silicates K_2SiO_4—rhombic and rhombohedral."[30]

Vernadsky proposed returning to St. Petersburg in December, partly to report on the progress of his research at a congress of scientists and physicians which Dokuchaev was organizing for January, partly to obtain permission from the Minister of Education to extend his stay in Western Europe for several months. "At the congress," he wrote Dokuchaev in October, "I will present a general theoretical sketch [of polymorphism]."[31] He arrived back at St. Petersburg in December, where he delivered his paper at the congress between Christmas and New Year's, arousing a lively discussion and apparently acquitting himself well. Vernadsky clearly enjoyed the opportunity to see many of his colleagues and to participate in scientific discussions. As he wrote his wife, who had remained in Paris, "I am feeling invigorated; in this respect the conference has been good for me. I feel that despite everything, science is the road for me."[32]

In the same letter he indicated that so far as job opportunities were concerned, he was exploring possibilities at Kiev, Odessa, and Kharkov. At the time, Kharkov seemed the best bet, although the possibility of a position at Moscow University soon opened up. The fact that the earth sciences were so underdeveloped in Russia at this time meant that Vernadsky could enter the field on the ground floor, the importance of which would grow exponentially as Russia began its rapid industrialization in the 1890s. During the summer of 1888, while attending an international geological congress in London, Vernadsky had made the acquaintance of A. P. Pavlov and his wife. Pavlov, about ten years his senior, was a professor of geology at Moscow, where he also taught mineralogy, a field he was eager to turn over to a younger colleague. Pavlov and his wife, a well known paleontologist, became better acquainted with Vernadsky on a tour of Wales organized for conference participants. They apparently were much impressed by the younger man. At any rate, later in 1889 Pavlov wrote Vernadsky inviting him to apply for a teaching position in mineralogy at Moscow University. During his visit to St. Petersburg in December and January, Vernadsky's advisor Dokuchaev, seemed well pleased with his student.

Vernadsky had begun to publish a number of articles and reviews in Western European and Russian scientific journals, and Vernadsky wrote to his wife that "Dokuchaev is making propaganda for me. Now he is praising my article on phosphorites as the 'best treatment in the Russian language.'"[33] Vernadsky was embarrassed by the praise, but also obviously pleased by the attention. Batalin, editor of official publications of the Ministry of State Domains (which included mines) had invited Vernadsky's participation in their publications, and Levinson-Lessing, a colleague from St. Petersburg University, had agreed to published an article of his entitled "Some Thoughts on Polymorphism" in the prestigious Scientific Herald (Vestnik estestvoznaniia).[34]

In contrast to the mood of these letters were Vernadsky's reports of several less exhilirating duties he had to perform while in the Russian capital. In a letter to his wife written in January of 1890, Vernadsky described the unpleasantness of his audience with Delianov, the minister of education, who apparently had to approve personally Vernadsky's petition to remain longer in Western Europe. It was Delianov who had tried to force Vernadsky's resignation as a graduate student at St. Petersburg University in 1887 on the grounds of "political unreliability." After his audience with Delianov in January of 1890, Vernadsky wrote his wife: "The conversation with Delianov was very superficial and produced a generally very unpleasant feeling, a feeling somehow of humiliation in the presence of such people who live among us and are upset by things foreign to us, repulsive to us, and at the same time they are so powerful that we must go to them, are dependent on them."[35]

While in St. Petersburg, Vernadsky also spent a good deal of time visiting relatives, both his wife's family and his own mother and two sisters. Although he always enjoyed staying with his in-laws, with whom he felt an intellectual and moral kinship, visits with his own mother and his sisters he found stressful. Since his father's death, his mother, who had never before been particularly religious, had turned more and more to what her son considered a conventional form of Russian Orthodox piety. He did not approve of this or of the various forms of health fads, such as homeopathy, which attracted her interest. He found he could not discuss ideas or social problems with his own relatives. Vernadsky remained very impatient with "small talk" and on this occasion commented that "we conversed about the usual things, more or less jumping from subject to subject, touching on reminiscences, relatives, acquaintances, thoughts about the future, about what we had seen . . . these conversations produce a strange impression . . . I felt the conversation constrained: those things which bothered or tormented the thoughts of everyone were not mentioned . . . and this could only succeed in giving the conversation an even more lively, impassioned tone. . . ."[36]

At his in-laws, Vernadsky felt more relaxed, despite his father-in-law's serious illness. The Staritskys were preparing to move to Poltava in the Ukraine where Egor Pavlovich would be on disability leave from his high bureaucratic position with the State Council. Vernadsky discussed with them his various job possibilities, which he felt now came down to a possible choice between the universities at Kharkov and Moscow. Until the offer from A. P. Pavlov in Moscow, Vernadsky had been thinking about applying for a position at the University of Kharkov in the Ukraine. A drier, warmer climate than St. Petersburg had been recommended for his wife, whose health had deteriorated since the birth of their son. Vernadsky's father-in-law was very much opposed to the Moscow possibility, basing his opposition on the colder, more humid climate of Moscow compared to the Ukraine. His daughter, he felt, would recover her health better in the Ukraine. Besides, Kharkov was only a four hour trip from the Staritsky's planned home in Poltava. Vernadsky respected his father-in-law's opinion but was still undecided when he left St. Petersburg. His mentor's opinion may have been decisive in the final decision, since Dokuchaev by May of 1890 strongly favored Vernadsky's applying for the Moscow position. The University of Moscow was stronger and better established than Kharkov. It was also closer to St. Petersburg, where Dokuchaev continued to work.[37]

Vernadsky returned to Paris in January of 1890 with much to think about and discuss with his wife. Their family life in the French capital kept to a strict regimen and conformed to the ascetic lifestyle which the members of their university circle in St. Petersburg had favored. The only luxuries they seem to have adopted were a French cook, who soon quit in disgust, telling the Vernadskys that she was accustomed to preparing a richer cuisine than the Vernadskys permitted, and an Estonian governess who accompanied Natasha from St. Petersburg to help care for their young son, Gulya (George Vernadsky), the future historian of Russia. One of the Russian friends they made during this period, V. E. Grabar, later a famous legal scholar, remembered in his memoirs that Natasha was little interested in housework (he may not have known that she was recovering from a serious illness) and devoted much of her time to their eighteen-month old son, writing all the details of his physical and mental development in a diary.[38] They rented a small house with a garden in a lower class suburb of Paris, more than a hour's ride by horse-drawn tram from where Vernadsky was working at various laboratories in the center of the city.

Vernadsky took advantage of the long commute to read widely in literature and philosophy. The experiments he was conducting also involved long periods of waiting for results, during which he continued his reading, devouring most of the ancient Greek philosophers, including Aristotle and Plato, whose ideas had been the subject of his father's doctoral dissertation many decades before. In addition, he read much contemporary European literature, histories such as Gibbon's on the decline and fall of the Roman Empire, and scientific works. Vernadsky browsed a great deal among the used bookstalls on the left bank, and looked, among other things, for the works of Russian radicals, which were banned in Russia.[39]

It was in Paris, for example, that he read the complete works of Alexander Herzen, considered by many the founder of Russian socialism. The publisher of Herzen's works was a Russian emigre scientist in Paris by the name of G. N. Vyrubov.[40] Vyrubov, as is turned out, was also a mineralogist whose acquaintance Vernadsky soon made. However, they quickly had a falling out. Apparently Vyrubov wrote a negative review for the *Bulletin of the French Mineralogical Society* concerning one of Vernadsky's recent articles in the *Zeitschrift für Kristallographie*.[41] Vernadsky felt betrayed, since Vyrubov had never mentioned to him in person his objections to the argument in Vernadsky's article. Vernadsky noted later with satisfaction that subsequent developments proved him correct and Vyrubov mistaken in the latter's objections. Generally, Vernadsky was not so sensitive to criticism, unless he considered it made in an underhanded way or unfairly. But in 1889 he may have been more than usually sensitive to such published critiques as he struggled to find a dissertation topic, establish a reputation as a mineralogist, and land a teaching job in Russia.

Besides working on his scientific topics, visiting the sights of Paris, and reading widely, Vernadsky also socialized a good deal among the Russian community living in that city. He and his wife liked having guests at their small suburban house. When possible they also liked to visit the apartment of the

Golshteins. Aleksandra Vasilevna Golshtein (née Bauler) was a political emigre who had left Russia when she was seventeen and settled in Paris, where she married another Russian emigre, a doctor who became editor of an important French medical journal. They were to remain lifelong friends of the Vernadskys and served as important intermediaries after the 1917 revolutions between members of the intelligentsia who remained in the Soviet Union and others who emigrated, including Vladimir and his son, George. The Golshteins frequently kept open house in their apartment on the Avenue Wagram, which became a gathering place for the liberal Russian intelligentsia in Paris. Here Vernadsky not only saw many of his friends from university days, such as the Oldenburg brothers and Ivan Grevs, he met V. E. Grabar and P. I. Novgorodtsev, later his close colleague at Moscow University. Novgorodtsev, together with Vernadsky and many of their friends, helped found the liberal Kadet party more than a decade later. A lecturer *(privat-dotsent)* at Moscow University, Novgorodtsev was later to become a well known professor of jurisprudence.[42]

Vernadsky's stay in Western Europe was important for several reasons: for the personal and scientific contacts made there, as well as for the intellectual direction he found in his scientific work, including his dissertation topic and other topics to which he devoted much of his time for the next decade. As already noted, the earth sciences were not yet well developed in Russia and tended until the 1890s to be largely descriptive and untheoretical, in contrast to the West.[43] If Vernadsky had not already been convinced, in Western Europe he realized the necessity for close contacts between Russian and Western science and cultural life in general. He was impressed by the progress of both theoretical and experimental sciences in the West and felt he learned there how to make more careful observations as a scientist and not rely so much on books and lectures for scientific knowledge, as he had in Russia. Western science was entering a period of revolution both in the development of instrumentation for more accurate observation and measurement and on the theoretical side. Vernadsky was to remain to the end of his life a strong advocate of close scientific ties with other countries, traveling abroad almost every summer in order to stay current with Western developments, until he was forbidden to do so by the Soviet government in the mid-1930s.

But his stay in Western Europe also increased his pride in the accomplishments of Russian science and its potential for the future, accentuated in part by what he perceived as the "scientific chauvinism" of many Westerners as well as by the growing recognition in the West that Russian scientists had a great deal to offer their colleagues in the West. The experience of overseeing Dokuchaev's exhibit of soil samples at the Paris Exposition and seeing his mentor win a gold medal there gave a strong boost to his pride in Russian science and his desire to publish his own work in the West when possible in order to receive recognition from the international scientific community.

Perhaps what was less typical for most Russian natural scientists at this time was Vernadsky's broad range of intellectual interests and desire to further his own education in philosophy, history, and the arts. In particular, his stay in the West strengthened his belief that ideas were the most important influence in human development and that ideas developed as the result of encouraging the emergence of free individual personalities. His personal philosophy, under the influence of Platonism and contemporary idealists such as his friends Driesch and Novgorodtsev, was to remain strongly idealist for the next few years, to the point where at times he doubted one's ability to know anything well outside the mind.[44] Through the influence of such idealism, he came to believe for a time in the immortality of the individual soul and the importance of having some kind of faith in the future as well as a desire for personal immortality which would spur the individual to greater accomplishments.[45] A non-churchgoer who remained critical of established religions and who was especially suspicious of state-dominated churches such as the Russian Orthodox, in Western Europe Vernadsky nonetheless continued his "religious" search for some kind of optimistic faith, which he believed necessary for the progress of human thought and human society.[46]

We cannot be sure exactly what sparked his interest in religious questions. What we can be sure of is that his religious ideas swung back and forth, for example between a belief and a disbelief in personal immortality, between an attraction for a pantheistic view of nature and skepticism about such a religious view in the decades of the 1880s and 1890s. For example, as a student in 1884, Vernadsky expressed his disbelief in the immortality of the individual soul. Then in Western Europe, he began to think more about this subject, and by the early 1890s, in a long conversation with the writer Lev Tolstoi, he expressed his faith in individual immortality, again abandoning this belief in later life. While in Western Europe in 1889 he saw such ambivalence as normal not only for himself but for human beings in general. Writing his wife after visiting an art museum in Germany, he noted:

> In essence you see here one and the same song; you see one and the same thought; you feel one and the same wish. This is mankind's idea of immortality, this desire to find satisfaction and an explanation of life and death. This song is about an ideal, about something better and higher than that which surrounds mankind at present . . . But doubtless the heaviest, the most tortuous, the most tragic thing in our life is the impossibility, in our mind and feelings, to come to peace with personal annihilation, with the absence of personal immortality. . . . Now scientific data, taken dispassionately, do not lead to anything that would answer this desire [for personal immortality], that would assuage this longing . . . I know and feel that this "longing" taken as a general phenomenon is not less strong than science. On which side victory will be, it is not possible to say.[47]

After the 1890s, Vernadsky was to abandon his belief in the immortality of the individual soul and his quasi-idealist epistemology for what he later described as "cosmic realism." He also came to believe more in the inseparability of the physical and the spiritual.[48] It is sufficient at this point to describe Vernadsky in Western Europe as a "seeker," interested in the development of a comprehensive worldview that would integrate all of human thought and experience, not simply the natural sciences, into a coherent point of view. His desire to avoid becoming a narrow specialist was certainly typical of the nineteenth century Russian intelligentsia, as was the seriousness of his search for some overarching meaning to life outside that of conventional institutions such as church and state. He remained highly suspicious of such institutions to the end of his life, considering them threats to the integrity of the individual personality, which he believed held the key to human progress.

Although much of the Russian intelligentsia in the late nineteenth century was attracted to socialism in one form or another, Vernadsky, most probably like the majority of Russian academic intelligentsia, remained highly dubious of socialist movements. While in France, however, Vernadsky not only read all of Herzen's works but also much of the work of French utopian socialists, whose ideas he found interesting even while remaining critical of socialism. In particular, he felt that socialism had little relevance for Russia at its current stage of development. As he wrote his friend Sergei Oldenburg in 1889, while still in Munich: "To introduce a socialist system now, immediately, is unthinkable . . . and as to what may be in two, three, generations, it is unreasonable for us to quarrel and argue about this now."[49]

However, Vernadsky's objections to socialism appeared to be based as much on its impracticality and untimeliness as it was on any objections to its collectivism or abolition of private property. "I consider this [socialism] more a matter of ideals than of practice," he continued in the same letter. Instead, what Vernadsky considered more practical was to struggle for representative, elective institutions in Russia and incremental reforms which would give more freedom to individuals and more voice in government to people outside the ruling circles. To dream about socialist revolution, he considered foolishness at Russia's present stage of development. He concluded his letter with the following statement, "As you know, I personally view 'socialist ideology' skeptically, in part because I am troubled about the fate of science and education should these people [socialists] be victorious and partly because the greater part of our land is inhabited by completely uncultured population, or a people uncultured in our sense of the word." He then added a strong note of nationalism, which for him was more important then socialism: "I want Russia, frankly, to be strong, mighty, and I think that she can do much, as can Asia, for the general development of Europe."[50]

Nonetheless, Vernadsky had already given some thought to what he would do should a social revolution, or some other form of major social change take place in Imperial Russia. In a letter written to his wife while he was still in Germany in 1888, he noted: "Only one means exists to prevent a decline of science and the arts in event of this or that possible social change (by whomever it may be

carried out, by the church, by one dynasty or another, or by a social revolution).
This is to work, with all one's strength and with all necessary sacrifices, to make
education more accessible and help it to permeate more deeply the masses of
people, not to stand aside from any social changes, not to go against them but to
connect them as tightly as possible with development of the arts and knowledge, to raise the means for acquiring knowledge (museums, laboratories,
universities) which will always be useful no matter what the cost. Still one more
thing—this is the widest possible diffusion of European and general cultural
influences in other parts of the world and an increase in exchanges among
various peoples."[51] Vernadsky displayed a simliar attitude at various later points
in his life, and this attitude did much to shape his relationship to the political
and social structures of Russia after 1917, striving to connect those changes as
much as possible with an increase in education, scientific research and the
general cultural level of Russia, despite his dislike of Marxism-Leninism and
the Stalinist bureaucratic centralism that eventually replaced the Tsarist regime. His decision to remain in Soviet Russia, which later puzzled and disappointed so many of his friends who chose to emigrate, including his own son,
was rooted in attitudes that were formed rather early in his career.

Vernadsky returned to Russia in July 1890 to participate in Dokuchaev's
expedition, investigating the soils of Poltava province, and to prepare for his
interview at Moscow University. As he wrote to his wife, who had remained in
Paris for the summer, he was tired of the nomadic existence they had been
living during the first four years of their marriage, and he was looking forward to
a settled life. He was particularly fearful that the Moscow job would go to
someone else, perhaps another student of Dokuchaev's or a candidate put
forward by the Ministry of Education, which under the 1884 charter had the
right to overrule a faculty's choice of a candidate and name someone preferred
by the ministry. Dokuchaev reassured Vernadsky that there was little competition for the Moscow job, an indication in itself of the dearth of young scientists
preparing themselves for careers in mineralogy.[52] Russian higher education in
the latter half of the nineteenth century had great difficulty preparing and
holding qualified young candidates for professorial positions, and many chairs
went unfilled in these years or were filled by candidates with lower qualifications.

Nonetheless, Vernadsky remained so concerned about his chances for the
Moscow position that he paid another visit to the minister of education, Delianov, to ascertain whether the ministry might put forward a candidate of its own.
Given his record of strained relations with Delianov and his reputation for
political unreliability among the bureaucratic elite, Vernadsky perhaps had
cause for concern. However, Dokuchaev put in a good word for him with the
ministry, and the young Russian scientist learned that the ministry did not
intend to sponsor a candidate of its own. The only remaining hurdle was to
impress the Moscow science faculty, which required the candidate to deliver

two public lectures as part of the interview process: one on a topic of the candidate's choosing and a second on a topic suggested by the faculty itself.

Vernadsky gave the first lecture, "On polymorphism as a general characteristic of matter," in late September. Although a perfectionist who was rarely satisfied with anything he did, Vernadsky received two rounds of applause from the large audience of students and faculty. Practically the entire natural sciences faculty attended this lecture and Vernadsky was personally congratulated afterwards by several senior professors, including his sponsor, A. P. Pavlov, and one of Russia's most noted biologists, K. A. Timiriazev.[53] In this lecture, which was subsequently published,[54] Vernadsky attempted to demonstrate the usefulness and applicability for crystallography of the ideas of a number of late nineteenth century Europe's leading scientists, several of whom, like Groth, Zonke, and Pierre Curie, Vernadsky had known and worked with during his recent stay in the West. In this lecture, he also attempted to apply to crystallography some of the ideas of the American mathematician, Gibbs, with whose work Le Chatelier had first made Vernadsky familiar and whose theory of phases Pierre Curie had attempted to apply in some of his work on crystals.

Vernadsky's second lecture, on a theme suggested by A. P. Pavlov and confirmed by the rest of the faculty, actually was closely related to Vernadsky's dissertation topic and was more specialized in nature. It dealt with new work on the synthesis of minerals, with which Vernadsky had become familiar while working in Fouqué's laboratory in Paris.[55] Although Vernadsky spent less time on this lecture and apparently wrote it at the last minute, it too went well and he was soon nominated unanimously by the faculty of Moscow University as a lecturer to teach mineralogy and crystallography, beginning in January of 1891.

In the meantime, he had three months to work on his dissertation and prepare his lectures, which had to be submitted for approval to the Ministry of Education. Vernadsky thus began a career which was to span exactly twenty years as a teacher at Moscow University, two of the most eventful decades in the history of Imperial Russia. These decades saw Vernadsky's scientific reputation grow as well as his participation in some of the most critical developments of Russia's political history, as a spokesman for the liberal academic community and the organs of local self-government, the zemstvos. These were also to be decades of great personal frustration on Vernadsky's part, as he felt himself continually tugged between work in a demanding specialty of science, broader intellectual interests which continually attracted him to a variety of other topics, especially the history and philosophy of science, and social pressures from his friends to participate actively in the movement for political and social reforms.

Vernadsky entered the Russian university system as a teacher at a particularly demoralizing time for the Russian professoriate. As a student, he had seen the effects during the 1880s of an increased centralization of power in the hands of the Ministry of Education and increased political repression. But it was only in

Moscow that he saw more clearly the negative effects of the University Statute of 1884 on the atmosphere for learning and scholarship. To dedicated teachers like A. P. Pavlov and others at Moscow University the arrival of this idealistic and dedicated young teacher, fresh from the capitals of Western science, was no doubt a breath of spring air, and Vernadsky was readily accepted as a colleague. The Pavlovs invited him frequently to dinners and open houses at their home (every Monday night was open house there). Even a senior scientist like I. M. Sechenov, a man in his early sixties already internationally known for his research on the physiology of the brain and other nerve centers and soon to be one of Russia's few Nobel Prize winners, took a strong liking to the young Vernadsky.[56] Sechenov began to drop by Vernadsky's laboratory to chat during his morning walks with a huge dog of undetermined breed. Vernadsky was obviously flattered by the attention and bothered only by the fact that Sechenov's dog once devoured his breakfast before he had a chance to complete it. Vernadsky soon had a wide circle of friends among Moscow's large academic community, laying the basis for his role as one of its chief spokesmen a decade later.

Vernadsky's major complaints about Moscow and the university there, upon first acquaintance, were the rather dusty provincialism of Russia's ancient capital, the sorry state of the facilities at Moscow University for research in his specialty, and the atmosphere of repression and corruption in the university, which he blamed on the government and the university statute of 1884. By comparison with St. Petersburg, Vernadsky found Moscow dirty and full of disagreeable odors. By contrast, even Poltava, the Ukrainian city from which he had just arrived after a summer of research, was a model of cleanliness.[57] The city itself, while it took some getting used to, was more tolerable than the situation at the university. Vernadsky's letters from this period are full of complaints. The university library lacked even the most basic texts in his field, the mineralogical collection had not been cleaned or catalogued since the 1850s, and laboratory equipment and teaching aids for lectures were scarce or non-existent.[58] All of this could be remedied but required long and exhausting expeditions through the maze of the university's bureaucracy. Less remediable, he soon found, was the fact that the Ministry of Education had decreed mineralogy and crystallography had to be taught in one year and according to a curriculum in conformity with the state examinations in these subjects. These were changes introduced as a result of the statute of 1884. Prior to 1884, mineralogy and crystallography had each been the subject of a separate year-long course, on the content of which students were tested in exams composed by their professors, not by the Ministry of Education in St. Petersburg.

Vernadsky considered one year woefully inadequate for fields which were rapidly changing. He felt that two years were really necessary in order to teach each subject adequately with attention to the latest scientific advances. The fact that he had to teach these subjects in too short a time and in order to prepare students for state examinations which were well behind the times in terms of scientific content made his job particularly frustrating.[59] "I clearly see and understand now the complete ignorance of the act of 1884," Vernadsky wrote to his wife in October.[60] In November he complained about a series of abuses: all power rested with the university administration and the procurator of the local educational district, who reported to St. Petersburg, rather than with the faculty, who were either demoralized in trying to do their jobs properly, or simply collected their salaries and neglected their lectures and students, keeping their jobs so long as they remained subservient to the administration. Underlings of the university administration openly accepted bribes for favors, space meant for science labs was being improperly used for housing, and an atmosphere of solidarity and participation in a worthwhile common endeavor among the faculty was largely lacking.[61] Even before he taught his first class at Moscow University, Vernadsky was already writing his wife that he was thinking about necessary reforms, a topic he was unable to do much about for nearly a decade. But it was the subject of university reform, about which he felt so strongly, that marked his debut as a political publicist in the national arena during 1902.

In 1890, the same year that Vernadsky was appointed a lecturer at Moscow University his former teacher at St. Petersburg University, D. I. Mendeleev, resigned after a severe reprimand by the minister of education, Delianov, for presenting a student petition to the minister. Little wonder that Vernadsky was nervous about the future of his university career when a world famous chemist such as Mendeleev could be reprimanded like a schoolboy.[62] Under the charter of 1884, the faculty councils of Russian universities were deprived of most of their powers. "Their functions were now mainly of a purely formal nature. The faculty council decided on student awards, conferred academic degrees, and passed on doctoral dissertations. The faculty council no longer even had the right to change the scheduled time of a particular class."[63] Gone were the days when faculty councils chose rectors and deans, appointed new faculty and controlled student discipline.

For Russian universities, the decade of the 1890s was marked by the further democratization of the student body as a declining percentage of students came from the ranks of the nobility and bureaucratic elite and more from the middle and lower middle class (as yet still very few from the peasantry or working class per se). This "democratization" of the student body also meant its increasing impoverishment. Financial aid for students during the 1880s and 1890s actually declined, and students were forced to rely more on aid from their families and their own earnings. Many were unable to make ends meet financially and an increasing proportion dropped out due to inability to pay student fees.[64]

Vernadsky began his university teaching career at a time when the percentage of students enrolling in the natural sciences was increasing. Paradoxical as it may seem in light of what has been said about university finances in general, the resources at the disposal of the natural science faculties were increasing, despite declining student financial aid and faculty salary levels, which had not increased since the 1870s. One of the basic problems for university scientists was a lack of adequate laboratory space, as the complaints not only of Vernadsky but of his colleagues at Moscow University, Sechenov and Timiriazev, indicate in the 1890s.[65] When Vernadsky began teaching in 1891, he complained not only of inadequate lab space but the fact that most of his students lacked familiarity with even the most basic laboratory equipment and techniques, a deficiency he quickly set out to remedy, with a good deal of success by the end of the 1890s.

The general atmosphere of suspicion and hostility between the universities and the government continued to increase during the 1890s. Student strikes and disturbances, which were met with police repression and mass expulsions, increased from 1896 on, with the advent of a new monarch, Nicholas II. The new Tsar was even more hostile toward the universities than his father, Alexander III. Nicholas had utter contempt for both faculty and students in the universities; and, under the influence of Count Witte and others, he continued his father's policy of expanding the network of specialized technical and commercial schools at the cost of stagnation for Russia's universities. In 1896, Vernadsky expressed great bitterness at this policy, opposing these new specialized higher technical schools on the grounds that their narrowly specialized curriculas were detrimental to the development of Russian science and technology. "The newspapers have carried an announcement of a new higher educational institution in Moscow—a specialized engineering school. Again we see the same situation—a narrow 'practical' curriculum, which of course will give neither practical experience nor produce an educated person. It is amazing how we have adopted and introduced into our life this creation of Napoleonic France—these so-called specialized schools. They will produce nothing, and in general I think money spent on them is thrown away, wasted."[66]

So far as the strained relations between the student body and the government were concerned, there were at least several sources of tension. One, already mentioned, was the impoverishment in which the majority of Russian university students lived. As one recent student of the universities in this period has stated, in the period after 1880, even as student enrollments increased, "the state made no effort to match the rise in student enrollments by a corresponding increase in its financial aid budget, and this was an important factor behind the increasingly strained relations between the state and the student population after 1884. Large numbers of students were forced to leave the university because they could not pay fees. In the first semester of the academic year 1899–1900, for example, 480 students, or twelve percent of the student body, were expelled by Moscow University for nonpayment of fees."[67]

At least equally important as a source of grievance was the regimentation and increasing repression of any independent student life. After 1884 the activities of the university inspectorate increased. It reported directly to the curator of the local educational district (an official of the Ministry of Education), thus bypassing not only the faculty but the local university administration as well.

The university inspectors worked closely with the police in spying on students and generally antagonized the student body by their arrogant and officious attitudes. In February 1901, the Faculty Council of Moscow University issued a report condemning many of the activities of the university inspectorate and blaming it for much of the student unrest of the previous years:

> The students are unanimous in their negative attitude towards this institution. . . . If a student, following the old and now dying tradition, enters the university at first with a feeling of reverence, then he is bound to experience some sharp disappointments almost immediately. First of all there are certain (noxious) formalities when he registers, then he must register his presence at the obligatory lectures. There are sub-inspectors who stand at the entrance of the lecture halls in order to ensure that no one not formally registered in the course enters the room. There are the sub-inspectors who patrol the corridors between lectures, negative marks received by students when their uniforms are not in order. At practically every step he takes, the students hear their comments, warnings, and remarks. . . . At first this all surprises the beginning student; later this lack of trust inflames and infuriates him, and this turns into fiery anger when the student sees that they are always following him, trying to ascertain whether or not he is "reliable." When the student is in the university he feels that he must be careful at all times, watch everything he says. . . .[68]

It was not unknown for university inspectors to burst into a student's room late at night, demanding information about other students suspected of illegal activities or to sit in student cafeterias and tea rooms all day trying to eavesdrop on conversation. Virtually all student clubs and organizations were prohibited, but some managed to exist, in a shadowy, illegal way. The most popular organizations in the 1890s were the so-called *zemliachestva*. As already discussed in connection with St. Petersburg University in the 1880s, these were organizations composed of students from the same provincial area, who were studying at the same university. They provided moral and financial support for their members and attracted considerable student membership. For example, there were forty-three such organizations at Moscow University in 1894, with some 1700 student members, representing more than a third of all the students at that institution. In 1896, the United Council of these organizations at Moscow University issued a proclamation explaining why students joined the *zemliachestva* and arguing that they should be legalized: "We have constant troubles with the police and cannot even leave Moscow without their permission. Instead of being regarded as the intellectual workers of the future, the Government insists on seeing us as 'separate visitors' of the university (with no right to any corporate life)."[69] The government met such appeals with repression, rather than legalization, leading to a wave of student unrest which grew steadily from 1896 until the revolution of 1905. Given this atmosphere and the uncertainties of a university teaching career, it is not surprising that Vernadsky considered what other alternatives might be open to him should his university career be terminated. In the decade of the 1890s, he spent a good deal of time and money refurbishing the estate he had inherited from his father in Tambov province, Vernadovka, building a modern home there for his family to live all year round if necessary and putting in order the economy of the estate so that he might be able to increase his income, particularly in the event that he and his family should someday become totally dependent on the earning of these 1450 acres of black soil. For example, in a letter to his wife in July of 1896, he discussed the possibility that the university might be closed in the fall due to student disturbances. Although he had once planned to sell the estate at Vernadovka, he now indicated that, given the situation at the university, it would be better to retain ownership of this land, improve it and prepare to live off its income if necessary. Given the political situation in Russia since Nicholas II's ascension to the throne the previous year, Vernadsky doubted that he would be able to sell the estate at Vernadovka, even if he so wished.[70]

Student disturbances in the second half of the 1890s were occasioned not only by distrust of the government and the way students were treated, but by the distrust of much of the faculty as well, whom most students viewed not as respected scholars and teachers, but as agents of the administration. Given this atmosphere, scarcely one conducive to learning, it is all the more remarkable that teachers like Vernadsky were able to build up a dedicated following among the student body. The contrast between such teachers, who basically loved their work and treated students with respect, and many other professors in the Russian university system who neglected their teaching responsibilities, undoubtedly increased the respect and even affection many students came to feel for professors like Vernadsky. Despite all the hardships of university life in these years, such teachers worked hard to advance the cause of learning and pass their knowledge on to students. As we will see later, this sense of affection and moral authority Vernadsky acquired among many students was as powerful a force in molding his future scientific school, which grew up first among his students at Moscow University, as was his very obvious talent as a scientific researcher. Moral authority and scientific competence went hand in hand in the creation of the Vernadsky school and helped to give it its particular force and cohesion in Russian science.

But it would be premature to speak of a "Vernadsky school" in Russian science until the later 1890s, when its first glimmerings can be found. In the years from 1891 until his appointment as an associate professor in 1898, Vernadsky was preoccupied with establishing himself as a respected scientist and teacher. It was during these years that he wrote two dissertations (the *magister* dissertation, completed in 1891, which assured the continuance of his appointment at Moscow University, and the doctoral dissertation, completed in 1897, which won him promotion from lecturer to a chair and a professorship in mineralogy). Vernadsky had intended to defend the *magister* dissertation in the spring of 1891, but he was late in completing it. It was based on three years of work, including a great deal of experimental evidence generated by Vernadsky himself as well as his reading of more than a thousand articles in the scientific literature. He had hoped to have it ready by January 1891, when he began teaching at Moscow University, but in March it was still not ready and he worked long hours trying to complete the dissertation while preparing six hours of new lectures each week, supervising lab work for his students, and helping his family settle in Moscow.[71]

By the time Vernadsky had prepared the required one hundred printed copies for the natural science faculty in St. Petersburg, Dokuchaev wrote back that he was too overburdened with other work to read it carefully and schedule a defense for the spring semester, despite the fact that other interested members of the faculty were agreeable to a late defense that spring.[72] As Dokuchaev explained to his student: "Your dissertation touches on such an important problem and refers to such a mass of [scientific] literature, that to *hurry* with its defense is simply *awkward* and not in your own personal interest."[73] Vernadsky's mentor was busy not only with his university duties but as a consultant to various government ministries and as a researcher analyzing the soils of various provinces, such as Poltava, for local zemstvos. In particular, he was an active member of a government commission appointed in 1891 to reorganize agricultural education in Russia. At this time. Russia contained only two higher educational institutions specializing in agriculture, and the government proposed closing one, the Novo-Aleksandriiskii Institute. This once venerable institution had fallen on hard times, and the commission, of which Dokuchaev was a member, hoped to reorganize it and revive its reputation. The commission's recommendation was accepted, and Dokuchaev was named its new rector in 1892, moving from St. Petersburg University, a fatal decision as it turned out.[74]

Dokuchaev's mind was obviously occupied with other matters in the spring of 1891 and the defense of Vernadsky's dissertation had to be postponed until the fall. Although Dokuchaev criticized the writing style and felt Vernadsky needed to improve his argumentation and way of expressing himself, by and large he was persuaded by Vernadsky's thesis and impressed by the quality of his scientific imagination. As he wrote his student in August after completing a careful reading of the dissertation: "Personally, I share the main argument of your work, with which, most likely, our chemists will also agree. I find your experiments and observations very interesting and valuable. In light of this I consider the dissertation completely satisfactory. Unfortunately, the exposition is not faultless and some of the factual material still needs fleshing out."[75] The defense was scheduled for October, and Vernadsky's two opponents were Dokuchaev and D. I. Konovalov, professor of chemistry at St. Petersburg University. Despite his anxieties, the defense, which was open to the public and attracted a large crowd, including some of Vernadsky's friends from his university days in St. Petersburg, went very well. Both the opponents raised general issues, rather than criticizing details, but no one found any fundamental flaws in his work. Afterwards, as Vernadsky wrote his wife, Konovalov was very kind and full of praise. Vernadsky planned to meet individually with everyone who had raised specific objections or points of criticism. From his own standpoint, he added that neither of his opponents was an expert in mineralogy, implying, perhaps too modestly, that he got off lightly.[76]

In his dissertation, Vernadsky refuted experimentally the previously held view that the aluminosilicates were "salts of silicic acid, and their acid properties" were attributable only to alumina. He "showed experimentally a different structure of aluminosilicates, according to which aluminum in the most important rock-forming minerals—feldspars and micas—is chemically analogous to silicon."[77] This was an important and well-founded contribution to mineralogy, but Vernadsky went beyond his experimental evidence to propose a theory of the chemical structure of the aluminosilicates, the most widespread group of minerals on earth. Vernadsky's interest in structural chemistry had been sparked by the teaching of A. M. Butlerov, one of the pioneers of structural chemistry and one his professors at St. Petersburg University. Vernadsky's theory was that of the kaolin nucleus, which he argued in his dissertation was shared by all the minerals of this family. In later work, Vernadsky estimated that eighty-five percent of the earth's core, to a depth of sixteen kilometers, consists of aluminosilicates, some indication of their importance.[78] This theory, which his French teacher, Le Chatelier, later dubbed "a brilliant hypothesis" in one of his textbooks of the 1920s, was not confirmed experimentally until the 1930s. Such verification came as a result of the development of X-ray structural analysis by Western European scientists, such as the German researcher Ernst Scheibold, who also acknowledged Vernadsky's "brilliant intuition" and confirmed its correctness experimentally in the early 1930s.[79]

A basic principle of organic chemistry is the existence of so-called radicals, groups of atoms which preserve their individuality in organic molecules and can pass from one type of organic molecule to another without changing their basic structure. To find such radicals in the minerals of the earth's core, however, was much more difficult because of the inability of chemists to use the methods appropriate to organic chemistry. All the same, Vernadsky was able to postulate correctly a fundamental radical in most aluminosilicates—the kaolin nucleus. With its help, Vernadsky united practically all aluminosilicates into a unified system.[80]

What Vernadsky proposed was that a special, complex, closed structure of atoms forms the crystalline basis of most aluminosilicates. The closed, circular structure of this nucleus is what assures its great stability.[81] The kaolin nucleus that his theory postulated was composed of two atoms of aluminum, two of silicon and seven of oxygen. "This theory has played an important role in explaining the structure, genesis, and classification of minerals . . . and it is now considered an established fact that silicon and aluminum in aluminosilicates are joined by atoms of oxygen placed at the points of tetrahedrons," as Vernadsky postulated in 1891.[82] He continued to develop his ideas on this subject and followed with great interest the experimental work in X-ray analysis of the structure of minerals later in his life, even as his major interests shifted to other areas of science in subsequent years.[83]

Vernadsky's mentor and his senior colleagues were very pleased with his scientific work and prospects for the future, and urged him to begin work on his second (doctoral) dissertation as soon as possible. Only with the doctoral degree in hand could he be promoted to the rank of professor and hold the chair in mineralogy at Moscow University, his ultimate goal. But before he could begin work on this new dissertation, a social crisis intervened, one which someone of Vernadsky's outlook could not ignore and one which soon swept him along in a wave of reviving social activism among members of Russia's liberal intelligentsia. Vernadsky completed the defense of his dissertation just as one of the worst famines in the history of nineteenth-century Russia, triggered by drought conditions in the spring and summer of 1891, was tightening its grip on large parts of the countryside, including Tambov province where Vernadsky's estate was located.

On October 3, 1891, only shortly before defending his dissertation, Vernadsky wrote A. I. Popov, the peasant who acted as overseer on his estate, requesting information about the conditions there. The reply he received was disturbing. Popov reported that he expected full famine conditions by mid-November. Already, many peasants were being forced to sell their livestock to the local gentry (*pomeshchiki*) at rock-bottom prices, and at least a quarter of the local peasants had begun baking "famine bread," that is, bread in which surrogates for rye flour were used, such as peas ground up and mixed with brick dust and hay.[84] According to Popov, families were already coming to Vernadovka to beg for help. Vernadsky saw the need to organize private relief work, since he shared the feeling of most of the critical intelligentsia that the efforts of the central government were inadequate and that the relief being distributed by the local zemstvos was insufficient and often unfairly distributed. Whereas many of the large absentee landowners made generous contributions to famine relief in Tambov province, the attitude of many of the smaller gentry who lived among the peasantry was one of apathy. Most considered the local peasants of low quality and not worthy of help. A frequent attitude displayed by such gentry was that the situation of the peasantry was caused by their own moral decline, connected with the decline of the traditional extended peasant family.[85] Such local gentry often played a decisive role in the organs of the local zemstvos, hence the feeling that relief from those bodies by itself was inadequate.

An American historian who has studied this famine and the relief measures undertaken by the central government has argued persuasively that the government responded rather effectively to the crisis. But this was clearly not the perception of members of the liberal intelligentsia at the time, whose strong biases against the Tsarist government may well have colored their own perceptions about the administration's effectiveness.[86] Whatever the truth may be, it is clear that the efforts of Vernadsky and his friends helped the local peasantry of Morshansk and Kirsanovsk townships in Tambov province weather the crisis much more easily than they would have been able to do without such private initiative. In the fall of 1891, Vernadsky asked his overseer to make a list of needy families in the areas adjacent to Vernadovka. The list came to 1,983 people. Although Vernadsky made frequent trips to Vernadovka during the fall and winter of 1891–92, he was unable to leave his duties at the university for long. Instead, he began to enlist his friends from university days in St. Petersburg, particularly members of the brotherhood they had formed there (the *Bratstvo Priiutino*). One of his closest friends, A. A. Kornilov, gave up his government job in the provinces to work full time in this famine relief effort. Two other friends, V. V. Keller and L. A. Obolianinov were in Tambov province by mid-December to organize famine relief kitchens. They first went to visit the famous writer Lev Tolstoi, to study the system of famine relief he was using, and then traveled to Vernadovka to use the same system there.[87]

In all, some thirteen friends and acquaintances of this group worked directly in famine relief in this region during the course of the crisis. Many others assisted through fund raising efforts. Such fund raising extended as far as Paris, where Ivan Grevs was staying. In France, he and the Golshteins raised over 6,000 rubles for famine relief by July 1892. Most remarkable of all perhaps was an anonymous donation of 30,000 rubles which the Vernadsky group received in January 1892, allowing them to increase their means four to five times. Vernadsky himself traveled to St. Petersburg to pick up the first 10,000 rubles of this anonymous donation on February 10, 1892.[88] The identity of the donor was not revealed during this period, but in the manuscript memoirs of Kornilov, written after 1917 and deposited in the Academy of Sciences archives in Moscow, the donor is revealed as Grand Duke Nikolai, uncle of the Tsar, who gave this money to the Vernadsky group only on the strict condition that his identity be kept secret.[89] By July 1892, when the crisis was over, these private efforts had resulted in opening 121 famine relief kitchens and feeding more than 6,000 people. Almost as important was the fact that they were also able to provide food for up to 1,000 horses, vital to the peasant economy, which might have died or been sold otherwise. The Vernadsky group also were able to buy horses for about 220 horseless families, a gift they presented, using a lottery system, on Easter Sunday 1892. Kornilov notes in his book about the famine, published in 1893, that in the ten years preceding the famine, many of these villages had lost up to one half of their livestock due to the harsh economic conditions of the time.[90]

The famine crisis brought many members of the liberal intelligentsia out of their drawing rooms, offices, laboratories or classrooms, to face the harsh realities of rural Russia, where over eighty percent of the population still lived. Although Vernadsky, for example, personally disliked much of the local officialdom and the parish clergy, whom he considered for the most part ignorant and parasitical,[91] he and his friends were able to enlist the active support of both elements, as well as village school teachers, for whom they felt a good deal more rapport. Priests and their wives, and local school teachers, took them around to the homes of the hungry in various villages and provided a great deal of necessary assistance in overcoming the hostility and suspicions of many peasants.

Some of the older peasant women at first thought that the help they were offered came from the Antichrist, who was seeking to buy their souls. Others thought that the help was a loan which everyone would have to repay later, so that everyone should receive relief equally, no matter what their need. This was an attitude strongly conditioned by centuries of experience in which the Tsarist government and nobility held the peasants collectively responsible for taxes, dues, debts, and army recruits. In overcoming such attitudes, therefore, the local clergy and school teachers, usually traditional enemies in the Russian villages, worked together. Most of the peasantry were finally convinced that the aid had no strings attached and that these young Russian nobles and "city-slickers" were doing good deeds for the sake of their souls and were not working for the government (or the devil).[92] This new movement "to the people" helped to revive the flagging spirits of the intelligentsia who had only recently lived through one of the most repressive and discouraging decades of the nineteenth century, the reaction of the 1880s. Kornilov ended his published report on their relief activities with the feeling that as a result of these months of effort to ward off famine (and the cholera and typhus epidemics that inevitably accompanied famine), a kind of spiritual tie had developed between the *intelligenty* and the peasantry.

Kornilov felt that the famine relief workers had come to understand the peasantry not abstractly but in their real situation. To what extent Kornilov was expressing the wishful thinking of the intelligentsia and to what degree such efforts really bridged the almost unfathomable gap in mentalities between these two groups is problematic. What can be said is that such private famine relief efforts, which were duplicated in many parts of the stricken countryside, were a major factor in the revival of social and political activism in opposition to the autocracy. Over the next decade, Vernadsky and his friends were to take an important role in such activism, leading eventually to the formation of the Constitutional Democratic party [Kadets] in which many of them, including Vernadsky were to play a vital part.[93]

For Vernadsky himself, the most immediate effect of the famine was a whirlwind of activity. In December 1891, he was made secretary of a special commission of the Moscow Committee on Literacy, a group of liberal intelligentsia frequently harassed by the Tsarist government. This commission was charged with providing help to elementary school children in famine areas.[94] In 1892, Vernadsky ran for and was elected a member of the local zemstvo in Tambov province, where he showed a special concern to widen educational opportunities for the local peasantry by expanding the system of elementary schools and raising pay for teachers. He was often discouraged by the apathy of the local gentry toward this work. For example, in 1892, in the wake of the famine and a decline in the local economy, most members of this zemstvo wanted to cut back on elementary schools. Vernadsky found himself the only member of the zemstvo board to speak out against the closing of schools in the face of a large deficit in tax receipts.[95]

In the same year, Vernadsky also found himself in a tangle with the Procurator of the Holy Synod of the Orthodox Church, Konstantin Pobedonostsev, one of the Tsar's closest advisors and one of those most hated by the intelligentsia. Pobedonostsev tried to prevent the opening, under the aegis of the local zemstvo, of secular Sunday schools in Morshansk township, where Vernadsky had his estate. Vernadsky sued Pobedonostsev in the Senate, the highest court in the Empire. Probably somewhat to his surprise, Vernadsky won the suit.[96] Most of Vernadsky's social activism, both in Moscow and Tambov province where he frequently travelled hundreds of miles by rail to attend zemstvo meetings, was related to education and the need to increase both its quality and opportunities for modern educations among the majority of the population. In the early 1890s, Vernadsky was thinking about a series of projects to further those aims, along with his zemstvo activity and work for the Moscow Committee on Literacy. A number of these projects never left the drawing boards; some were realized many years later; all were significant for what they reveal of his approach to a peaceful "cultural revolution" in Russia, which, together with his work to expand the frontiers of scientific knowledge, was the overarching project of his lifetime.

One of these projects, briefly attempted and then abandoned for lack of financial support, was to publish a magazine to spread the political views of his group. According to one report, as early as 1890, while still studying in Western Europe, Vernadsky had been part of a group which included A. A. Kornilov, S. F. Oldenburg, and several other university friends who purchased the journal *Severnyi vestnik* for this purpose.[97] Although this venture failed, Vernadsky was thinking again in the spring of 1892 of starting a journal of self-education for the newly literate.[98] In another letter a month later, he spoke of promoting a series of popular biographies of people who had fought for freedom of belief. This, he believed, would be a good way to spread constructive ideals among the people.[99] In the early 1890s he was also thinking of teaching a course and writing a book on the development of scientific thinking, as a means of disseminating a more modern worldview in Russia.[100] The latter project was eventually realized, but not for more than another decade.

In 1893 and 1894, Vernadsky also participated in several illegal conferences of zemstvo activists. "At these meetings, discussion of matters directly related to the famine and the agrarian crisis gradually gave way to discussion of more general zemstvo problems, and then to critiques of the zemstvo statutes of 1890, which limited their functions and tax base, the institution of the land captains *(zemskie nachalniki)* introduced by the legislation of 1889, and other aspects of the 'counterreforms' of the immediately preceding years."[101] It probably did not help Vernadsky to learn that he was under sporadic police surveillance for his "subversive" activities, a fact blurted out by the wife of the minister of interior to his father-in-law, during a chance encounter and later confirmed by government documents found after the 1917 revolutions.[102] As one of these documents in the police archives read: "In the beginning of the 1890s, he moved from Petersburg to live in Moscow, where he continued his dubious acquaintanceships, took an active part in evenings organized by students of Moscow University where he gave speeches about the necessity of coming together of professors and students for purposes of political education of youth and struggle with the present regime."[103] In the 1890s, Vernadsky began to notice middle-aged men in bowler hats near his home every morning when he left for the university. At first, mistaking them for neighbors, he bowed politely in recognition. Then one day, preparing to leave for Western Europe, he noticed one of these same men watching him at the Moscow train station. Vernadsky mentioned this later to the liberal zemstvo activist I. I. Petrunkevich, one of his closest Moscow friends. Petrunkevich explained that they were probably all under surveillance by the *Okhrana*. Petrunkevich had himself attracted the "Tsar's eye" (*Tsarskoe oko*) many years before for his constitutionalist aspirations and had been under sporadic surveillance ever since.[104]

The Moscow scientist's life became so filled with political and social activities he complained that he could not concentrate on his scientific work. In the spring of 1892, for example, he had promised Dokuchaev a long overdue report about his research during the summer of 1891, on the soils of Poltava province. Dokuchaev, in a series of letters, begged and cajoled Vernadsky for the report, complaining that Vernadsky's slowness was "stopping the whole machine."[105] Vernadsky wrote to his wife about a sense of malaise in pursuing his scientific work, which apparently had overcome him since the completion of his dissertation. For example, at the same time that he was supposed to be writing the report for Dokuchaev, in March of 1892, he noted in a letter to his wife that he had written 107 letters in a period of 33 days to a variety of friends and acquaintances.[106] Besides Dokuchaev's report and the doctoral dissertation, which hung over him like a sword of Damocles, Vernadsky had promised to write reviews of Russian works in crystallography for Professor Groth's journal in Germany. Vernadsky reneged on his promise in 1892, a decision which very much annoyed the German scientist, but did not permanently disrupt their ties.[107] Natasha, Vernadsky's wife, summed up the situation very well in a letter to their friends, the Golshteins, in Paris: "In Russia, it is difficult to be a scientist. The surrounding life swallows one up."[108]

After several false starts, Vernadsky finally completed work on his doctoral dissertation in 1896 and received the degree the following year, defending his thesis at Moscow University. Although he originally intended to write on the subject of polymorphism in minerals, which had fascinated him for years, he decided that the subject was too broad and would require too many years of experimental work to do properly. He chose instead a narrower topic in crystallography, in order to have the degree out of the way and qualify for a professorship.[109]

Vernadsky's work in crystallography in general was concerned with the relationship of crystal form to the underlying "physiochemical structure and emphasized the importance of energetics in studying crystals."[110] The particular problem he chose for his doctoral dissertation was the phenomena of gliding in crystal substances. As he explained it to his wife while completing most of the work in the summer of 1896, the dissertation would contain five chapters and a conclusion, the first a general introduction explaining the significance of the problem and discussing previous literature, the following three on the phenomena of gliding in rock salt, *izvestkovoi shpat* (spar), and several other minerals, giving a summary of known instances of these phenomena, the fifth chapter would be on the connection between the planes involved in gliding and the symmetry of crystals. Vernadsky's intention was to proceed from the particular, his own experimental observations and measurement of the phenomena, to some general conclusions at the end.[111] He completed much of the work by the beginning of August (over two hundred pages), but remained unhappy with it as he left Russia to join his wife in Paris for the remainder of the summer:

> I cannot say that I am satisfied with the work. It is reminiscent of a bear's dance, that is, it can scarcely be considered on a high level esthetically. But from the standpoint of esthetics, sometimes, individual leaps of the bear are successful. My work represents such a bear's dance in science. Most of all I am vexed by the fact that, due to malaise, there are a whole series of questions which I have not treated. I am too lazy to do complex measurements, too lazy to carry out proper mathematical preparation; and now, preparing this work for printing, I feel a whole series of fundamental questions I have not treated. I will return to some of these questions, but in general I have the feeling that I have approached and skirted, come close to discovery, but have made none, due to laziness and dilettantism.[112]

Vernadsky tended at times to be very harsh in his judgements of himself. But it is clear that in the summer of 1896 he was occupied with many other topics, even taking time out to read the novels of Sir Walter Scott, along with historical accounts of English pirates and other "recreational reading" which in part constituted his self-professed "dilettantism." More serious, however, were his persistent worries about the general political situation and its possible effects on the university. Vernadsky complained of the repressive and archreactionary environment among many elements in the government and nobility, which was reflected in severe restrictions being placed on the private initiative of social service organizations such as the Moscow Committee on Literacy.[113] "Everywhere one meets a kind of nervous, tense, angry mood and it is difficult to concentrate on scientific work that is far removed from this earthly hurly-burly."[114]

Also that spring and summer dozens of students were expelled from Moscow University for participation in illegal student organizations such as the United

Council of Zemliachestvos. A physician and teacher in the university's medical school, Friedrich Erismann, was arbitrarily and illegally forced out of his teaching job that summer while he was abroad, supposedly for having ties with the United Council. All of this Vernadsky found very troubling. "They say that the decision came not from the university administration but from the Grand Duke, Sergei Aleksandrovich. . . . However it may have occurred, this is an act of offensive force and complete arbitrariness."[115] Although by September most of the expelled students had been reinstated, Vernadsky felt a great sense of foreboding about Russia's immediate political future and particularly the situation of the universities. "The year threatens to be difficult, oppressive, all the more so since the mood of society is now completely different than it was several years ago and I foresee with some fear and sadness a frightening near future, where reactionary forces, who understand nothing, are leading Russia."[116] Erismann, the dismissed university teacher, was given only a few days notice to resign, according to Vernadsky; and when he refused, he was fired. "He is very depressed, angry, and understandably insulted. His resignation was proposed to him *without any explanation of the reasons*." Vernadsky indicated that Erismann had already been replaced by a man named Bubnov, "a very bad person and in no way a qualified scientist."[117]

Erismann, a Swiss citizen, married one of Russia's first women physicians, whom he met at the medical school in Zurich. In the 1880s he returned to Russia with her and became one of the important pioneers in the public health movement and in the Pirogov Society, a professional society of publicly employed physicians in the Russian Empire. As an active and outspoken participant in the Moscow zemstvo, he had also become a thorn in the side of the Moscow governor-general and provincial administration. After his dismissal from Moscow University, he never returned to the Russian Empire, but he became a kind of martyr in the eyes of the liberal intelligentsia and especially to public health physicians.[118]

Whereas in one sense the social environment was distracting for Vernadsky's scientific research, in another sense he found in science a kind of refuge from the developing turmoil of Russian society at this time: "Every such incident forces one to work all the harder, one wants to accomplish more, to think more intensely. One wants to be closer to what is eternal in the scientific quest and to find there tranquility and a feeling of independence . . ."[119] The sense of internal emigration, which later became strong for many Soviet scientists and other intellectuals, was already present as a cultural force for Russian thinkers like Vernadsky, as it undoubtedly had been for many of his predecessors during the Tsarist period.

Underlying all of this, Vernadsky indicated in October, was an internal conflict which was not unusual for members of the Russian intelligentsia in these years, a resistance to becoming a narrow specialist and a desire to remain a generalist in science and scholarship, broadly informed intellectually, as well as socially and morally involved in the affairs of the world:

I feel that I am becoming a specialist, part of my interests are receding and although along with this the intensity of work in a specialized area is strengthened this is connected with a well known narrowing of the mind. In this respect my expedition to the Urals [which he was then visiting with his students] has done much for me and two roads have clearly opened before me—one, although little productive and partly dillettantism, at the same time forces the mind to work more intensely and more broadly, the other is more productive, more defined—but at the same time confines the mind within specific parameters and inevitably shrinks the horizon, placing a person within the ranks of scientific workers but not among the creators of the unfolding process. My attitude toward this latter road is ambivalent. If everything goes normally, I will doubtless choose the second road from a sense of duty. Only certain select minds are capable of combining these two sides of scientific thought, and I do not belong among them. I will be a good worker in a more or less narrow scientific area.[120]

However, as events were to prove, Vernadsky was to swing widely from one extreme to the other, at times doing solid, specialized field work and experimental work and at other times making broad generalizations, even creative speculation in a number of areas on the frontiers of science, which provided direction for the more concrete and specialized work of others. The result was that he left many projects uncompleted or only partially completed and jumped about from topic to topic a great deal during his career. One might see him as a transitional figure in the history of the Russian intelligentsia. He stood on the boundaries between the intelligentsia's broadly generalist past when Russian thinkers took pride in being able to write and converse on a wide range of topics but often accomplished little that was concrete and future *intelligenty*, particularly under the Soviet system, when the ordinary member of this group was expected to be a *spets*, a specialist in a particular, often rather narrow area of learning. Psychologically, for Vernadsky, as probably for other members of his generation, this transition created a tension and continuing sense of ambivalence, uncertainty, and even guilt about his social role. First, there was the tension felt within science itself between being a specialist and a generalist; second, there was a tension between science as a profession and the pull and tug of social activism. On the latter score, Vernadsky wrote his wife in July: "Science will always be more important than social service." Unlike some of his friends from St. Petersburg University days, the latter area he always felt as a hard duty, and was happier when he could return to the world of scholarship.[121]

Although Vernadsky tended at times to be a very harsh self critic, his doctoral dissertation was approved with praise by his university and judged a creative piece of work by a later generation of Russian specialists in crystallography. As one of the foremost Soviet crystallographers, Shafronovsky, wrote in later years:

Here we find the richest synthesis of data relating to unique deformations of crystals, created as a result of gliding, that is the shifting of separate parts of a crystal along straight lines while preserving the volume, weight, and homogeneity of matter. Vernadsky revealed the connection between the planes of gliding, the crystalline facets and elements of symmetry. Here for the first time, he underlined the need to make several qualifications in our conceptions about the complete homogeneity of crystalline polyhedrons in connection with changes in their physical features in their surface state. According to this idea, crystals are viewed not as abstract geometrical systems, but as real physical bodies. At the present time, the proposition put forward by Vernadsky is generally accepted.[122]

Just before his defense, Vernadsky submitted a copy of his dissertation for comments to the man who was probably the most creative and best known crystallographer in Russia at the time,. E. S. Fedorov, a professor of geology at the Moscow Agricultural Institute. In his response, Fedorov was both complimentary and critical of Vernadsky's work. As he wrote on March 13, 1897:

> I want to congratulate you on a very interesting and instructive piece of research, "On the Planes of Gliding," and thank you sincerely for sending me this work, from which I have extracted much that is useful, although I have not had time to read it through as carefully as it deserved. I regret that the work came into my hands just as my textbook on crystallography was being printed. Otherwise I would have used material from it that is instructive.[123]

Fedorov went on to criticize Vernadsky's use of the term "gliding" (*skolzhenie*) arguing that what he had investigated in his thesis was not gliding at all but another set of phenomena found in crystals, deformations he called displacement (*sdvig*). This disagreement did not prevent Fedorov from giving the younger Russian scientist high praise when he reviewed the published version of Vernadsky's doctoral dissertation of 1898. Comparing Vernadsky's dissertation with the work of three other contemporary scientists who dealt with various anomalies and deformations in crystals, Fedorov found the work of Vernadsky and one other of these authors to be superior to the other two, since they were distinguished by their completeness and their originality. He was particularly impressed by Vernadsky's critical attitude toward preceding work, his many original observations, and his independent views. Developing some of the critical views he had earlier expressed to the author privately in the letter cited above, he nonetheless closed his review with high praise: "If I elaborated on the weak sides of this work, then it is exclusively because I give this work very great importance and I would be very happy to see these weak sides ironed out in later works by this highly promising author."[124]

Vernadsky continued an interest in crystallography until the end of his teaching career at Moscow University in 1911 and published his first major scientific book, *The Fundamentals of Crystallography*, in 1903. However, it was not as a crystallographer that he was to become well known as a scientist; and during the course of a career in which he published over 400 articles and books, only about seven percent was devoted to the study of crystals per se.[125]

Nonetheless, his work in this area is significant for several reasons. Vernadsky's historical and analytical approach to crystallography was different from most of his contemporaries' descriptive studies. For most crystallographers at the beginning of the twentieth century, such as Fedorov, their science was primarily geometrical. By 1900, using mathematics and a variety of special scientific instruments, they had reduced all known crystals to thirty-two classes and two hundred thirty different groups, based on their geometrical form.

Vernadsky's own interests went beyond the descriptive and mathematical. In his *Fundamentals of Crystallography*,[126] he asked why crystals have such special geometric forms. For him the study of crystals was interesting for what it could reveal concerning the structure of matter in one of its three fundamental states, solids (as contrasted with liquids and gases), and he sought to connect the form of crystals with the underlying molecular and atomic structures they reflect.[127] For Vernadsky, "crystallography is concerned with the study of the laws of the solid state of matter."[128] In his book, he gave a short historical survey of the development of the field and noted that traditionally the study of crystals had developed independently of physics and chemistry. He sought to bring these subjects together in a closer marriage. He viewed crystallography as a subdivision of mineralogy, since all crystals are, at the same time, minerals. "It is impossible to be a mineralogist," he observed, "without mastering the basic methods of crystallography . . . because the mineralogist is concerned with the solid, crystalline products of chemical reactions on earth."[129]

Throughout his career, Vernadsky was to remain interested in the different forms that matter takes in space, and the reasons for those differences, for example between the geometric shapes that living matter takes as compared with inert matter, which he came to view as fundamentally different states of matter. He posed the question why this should be, a question which had also been posed by other scientists before him, such as Pierre Curie. It was a fundamental question which was never adequately answered by any of them. But it was crucial to focus the attention of science on the different spatial forms that matter takes and pose the question, and Vernadsky participated in this process. Physicists at the time Vernadsky was writing about crystallography, in the early 1900s, studied the movement of matter in an "inert, structureless medium (space or the ether)." In contrast, Vernadsky's interest was consonant with an important direction in science, the relativistic universe to be posited by Einstein in these same years, the view of the universe in its constituent parts as a structured, active medium in which matter can assume many forms in space, dependent on a variety of factors, including its historical development.[130]

Vernadsky's first major book was indicative of his approach to science. In several respects it went beyond description to pose the question why, suggesting answers based not only on structure at a particular moment in time but on historical development. It looked for general laws in all the particular phenomena of nature; it included a thorough survey of the existing literature and an historical sketch of the history of ideas in the field; and it sought to find the connections between one particular branch of science and other fields. Finally, it was rich not only in suggesting possible answers to the questions posed but in posing new questions and problems for research.[131] Thus, Vernadsky's book on crystallography came to be considered a classic of its kind in the Russian literature and was read with profit by following generations of scientists interested in this field.[132]

Shortly after the publication of this monograph, Vernadsky returned to a far more ambitious project in mineralogy, for which he had begun gathering material in the 1890s. This was a kind of encyclopedia of the "mineral kingdom" of the Russian Empire. His aim was to gather together and systematically analyze all the known data about the genesis, characteristics, deposits, and industrial significance of all minerals then known in Russia.[133] This work was entitled *An Attempt at a Descriptive Mineralogy*, although the title was a bit misleading, since he obviously hoped to do more than describe minerals in the usual sense. In this work, he wanted to analyze the historical development, structure, known deposits, and uses of every mineral found in Russia. This was obviously more than a single person could hope to accomplish in a lifetime, and Vernadsky began to enlist the help of both his students and other scientists.

As Vernadsky later noted, he received the help of everyone he turned to in Russia, with one exception, the crystallographer E. S. Fedorov, who had become rector of the Mining Institute in St. Petersburg and answered Vernadsky's plea for help with a sharp criticism of the project, which he misunderstood as a dry collection of facts. He considered it impossible ever to complete the project and stated as much to Vernadsky.[134] In this respect, Fedorov was probably correct, since Vernadsky did not complete the project, although what he published proved to be valuable. In another respect, however, his sharp note of rebuke was unfair, since he accused Vernadsky of undertaking the project with purely commercial aims in mind rather than a scientific purpose. There is no evidence that Vernadsky's motives here were commercial, or that he ever realized a profit from those portions of the *Attempt at a Descriptive Mineralogy*, which were finally published.[135] Vernadsky had major ambitions which he often did not fully realize, but a desire to make a large profit does not seem to have been among them. Recognition, even fame, were certainly among the motives that emerge from a careful examination of Vernadsky's legacy, but a strong desire for monetary profit seems much less evident and is contradicted by Vernadsky's own lifestyle, which was ascetic, comfortable but far from luxurious.

In the relationship between Vernadsky and Fedorov there was an undercurrent of competition and at times resentment, at least on the part of Fedorov, who had a very difficult time achieving recognition in his native land and did not obtain a permanent academic position until the age of forty, much later than Vernadsky. Fedorov came from a bureaucratic family which had fallen on hard times during his teenage years in 1866, after the death of his father, a major general in the engineering corps who had worked his way up from the peasant estate.[136] As a student in the Military Engineering School in St. Petersburg, Fedorov first became involved in illegal political activity in 1869. After two years of military service as a second lieutenant, he resigned from the service in 1874, and by 1876 he was deeply involved in revolutionary activity as a member of the populist Land and Freedom Movement *(Zemlia i volia)*. Fedorov hid a revolutionary printing press in his apartment in St. Petersburg and was commissioned by other Russian revolutionaries to establish ties with revolutionary groups in Western Europe. Although he later dropped his revolutionary connections, it was no doubt in large part due to this past that his scientific career met stumbling blocks in Russia.

In the early 1890s, Fedorov had been proposed for membership in the Imperial Academy of Sciences, on the basis of his mathematical studies of crystals, but he was voted down. Thereafter, he was forced to work as a mine geologist in a distant province for some years. It was ony after his work was recognized and praised in the West that Fedorov was able to obtain a permanent professorship in the later 1890s, becoming a professor of geology at the Moscow Agricultural Institute in 1895 and then director of the Mining Academy in St. Petersburg in 1905. He was the first rector elected by the faculty of that Academy, after they received the right to elect their own deans and rectors during the revolution of 1905.[137]

By comparison, it must have seemed to the older crystallographer that Vernadsky, who foreswore underground revolutionary activity, who believed in peaceful evolution and was from the hereditary nobility, raised in comfortable circumstances, had found his academic path much smoother and recognition much more easily obtained. It is ironic that when a vacancy in mineralogy occurred in the Imperial Academy of Sciences in 1905, Fedorov would not permit his name to be proposed for a second time. To Vernadsky's great surprise, therefore, he was proposed to fill this vacant slot by his friend and former university associate, Sergei Oldenburg, who had become permanent secretary of the Academy. Fedorov, whose accomplishments at the time were considered more significant, was bypassed once again for membership in the Academy, and Vernadsky was elected as a adjunct member with ease, achieving full membership in 1912. To Vernadsky's credit, he always spoke highly of Fedorov in print; and it was in large part due to Vernadsky's detailed scientific recommendation that Fedorov finally was elected a full member of the Academy in 1918, following the Russian Revolution.[138]

Although the two men's careers had converged briefly, they actually represented quite different tendencies in Russian science. Fedorov was a brilliant specialist, whose original work was largely concentrated in the area of mathematical crystallography, although he later dabbled briefly in the philosophy of science. Vernadsky, as we have seen, had a far ranging philosophical and analytical mind with a preference for generalization, synthesis, not to mention speculation at the crossroads between a number of specialities. Although at times a brilliant experimentalist, Vernadsky's interests obviously lay elsewhere and he had difficulty staying put in any single specialized area of science.

Nonetheless, in the 1890s most of Vernadsky's scientific work was concentrated in the fields he taught at Moscow University: crystallography and mineralogy. By the mid-1890s he had begun to attract a number of talented students who were to carry on his work in these fields long after his own interests had largely shifted. In 1903, at the age of forty, he did a summing up for himself and his friends.[139] Despite a great deal of agonizing in the previous years about his dilletantism and inability to accomplish all he had hoped to accomplish, he took pride in creating a mineralogical laboratory at Moscow University which he considered the equal of any in the West.[140] He had carefully chosen those students he wished to work with closely; and by 1903, 10–15 undergraduates and graduates worked in his lab on a regular basis. Although never known as a brilliant lecturer (his lectures had a reputation for being dry at times, and he tended to become sidetracked, sometimes speaking over the heads of students),[141] he made up for such defects by his reputation for rigorous logic, enormous erudition, careful supervision of field and lab work, and perhaps most important of all for many students, concern and respect for them as human beings. He also had a reputation for being enormously demanding of his students and inspired not only respect but a certain amount of anxiety on the part of students who feared being unable to please him.[142] Before accepting any student in his laboratory, he and his chief laboratory assistant, a graduate student by the name of P. K. Aleksat, would subject the prospective mineralogist to a rigorous interview and series of tests to gauge their knowledge and ability for the field as well as their motivation.[143] Aleksandr Fersman, one of Vernadsky's most famous pupils and one of his favorites, still remembered in 1945 the awe and fear he experienced when he walked into Vernadsky's lab as an undergraduate one day in 1903, asking to be accepted as a student in mineralogy:

> I was so anxious that I could hardly speak; and the professor [Vernadsky], scrutinizing me through his thick glasses, seemed to me very strict. He directed me to a small room of twelve square meters—the mineralogical laboratory—to his yet more frightening assistant. They showed me to a place in the corner near the stove and gave me a mineral to study: yarozite from the Island of Cheleken. Thus began my long association with V. I. Vernadsky and his remarkable assistant who perished so tragically, Aleksat. Thus began the five remarkable years of my university life in Moscow, in the friendly family of mineralogists.[144]

The searching eyes of Vernadsky, which had at first seemed so severe, turned out to be merely sizing up this prospective student, as the professor and assistant probed Fersman for the extent of his knowledge and interest in the subject. Mutual respect and affection soon developed between Vernadsky and

Fersman which was to last a lifetime, as it did with so many of Vernadsky's students. As Fersman wrote soon after his mentor's death in 1945:

> His wonderful figure still stands before me: the quiet, unaffected, scholarly thinker; his splendid, clear, sometimes joyous, sometimes thoughtful, always beautiful, radiant eyes; his slightly nervous, quick manner, the handsome grey head of the teacher; the figure of a man of rare purity and beauty, which manifested itself in every work, every movement and action.[145]

Such devotion was not won easily from a skeptical student population who viewed so many of their professors with suspicion or scarcely veiled contempt, as servitors of a regime which most students despised. In Vernadsky's case, he earned it by his obvious talent as a scientist and even more so by his personal qualities of devotion to his students and their education.

Throughout his life, Vernadsky always had a few favorite students in whom he confided and to whom he entrusted extra responsibility. Fersman turned out to be one of his favorites, although he was in many ways the polar opposite of Vernadsky: the professor of a very concrete, factual mind, unspeculative, unphilosophical and apolitical. (Fersman's father at the time was a Tsarist general, newly appointed commander of the Moscow Second Cadet Corps, and Fersman deliberately remained aloof from student politics of whatever brand.) Fersman was also less moody and serious than Vernadsky in some respects, something of a practical joker, who lightened the mood of any room he entered. As a member of Vernadsky's household later reminisced, the middle-aged professor brightened up whenever Fersman came to visit. The servants at the Vernadsky household took to calling the chubby, already balding young aristocrat in his early twenties, "our wind-up baron," since he reminded them of the humorous windup toys children receive at Christmas.[146] While his personality was no doubt attractive, what won Vernadsky's strongest affection was Fersman's enormous capacity for work, a mind quick to grasp new concepts, and a total dedication to developing the natural sciences in Russia, particularly those to which Vernadsky had dedicated his life.

Vernadsky was always on the lookout for such talent, especially since by 1903, at the age of forty, he was looking toward the final years of his life and the need to find successors who could carry on the work. He had so many interests and projects which he knew he could never personally complete that the idea of a "collective" of younger scientists increasingly appealed to him. As he told his wife in 1903, the large-scale project to create a mineralogy of the Russian Empire which he was then planning, he viewed as a collective work, to be completed with help from a number of his students.

On the one hand, he was enthusiastic about this new endeavor, which he thought would require five to six years for completion. On the other hand, he knew it would necessitate a great deal of field work which would take him away from his lab and another project he now wished to undertake: an investigation of radioactive minerals and their significance for the field of mineralogy. Since Becquerel's discovery of radioactivity in 1896 and the experiments of Pierre and Marie Curie, Vernadsky had become fascinated by the implications of these developments for the earth sciences. He felt himself present at the beginning of a scientific revolution which would change science's understanding of nature. It could also mean great recognition for scientists who understood this revolution. The university's physicists had promised to provide him with some radium and polonium, and he was anxious to start this new work.[147]

Since coming to Moscow University in 1890, Vernadsky had carefully assembled three collections vital to his work: a collection of rare minerals, one of the finest in Europe, a collection of scientific instruments painstakingly assembled over more than a decade from the scarce resources of the university. The third collection, however, was the most valuable and fragile: a group of people with both the intelligence and ability to work together harmoniously. Over the years, he was to lose a number of them, some to tuberculosis, some to suicide, some to war and revolutionary violence, others to hard necessity when they were forced to leave scientific research and work in industry or the bureaucracy because of financial hardship. In the conditions of the Russian Empire at the beginning of the twentieth century, it was not easy to create and maintain a talented group of people in science. The losses began in the 1890s and became especially painful from 1902 on. In that year Vernadsky's pupil and lab assistant since 1895, a quiet and hard-working Ukrainian by the name of Anatolii Orestovich Shkliarevsky, succumbed to tuberculosis of the throat, a condition which some of Vernadsky's students felt was aggravated by working in the dusty Mineralogical Museum and the damp chemical filled air of the small basement laboratory which Vernadsky maintained at the university.[148]

It was a tribute to Vernadsky's interpersonal skills that such a diverse group of people were able to cooperate in a common endeavor, since they included not only Russians from a variety of backgrounds, including aristocrats like Fersman who was of mixed German-Russian background, but Poles, Jews, Ukrainians, and several talented young women from the Higher Women's Courses where Vernadsky had begun to teach in 1897. Women were still excluded from Russian universities, but were allowed to attend a separate institution supported by private donations where most of the liberal professoriate taught in this period for no extra compensation. University authorities looked aside when Vernadsky began to allow women to work, illegally, in his laboratory at the University. Several of the women he recruited in this way became professional mineralogists and geologists as the result of their early work with Vernadsky. At least one, E. D. Revutskaia, worked as a scientist in Vernadsky's lab for much of the remainder of his life. She clearly revered her teacher.[149]

One of Vernadsky's most talented followers in these years was Iakov Vladimirovich Samoilov, who came to Moscow in the mid-1890s after receiving his bachelor's degree in the natural sciences from Novorossiisk University in the south of Russia. A small, rather frail young man, Samoilov impressed Vernadsky with his great energy and desire to master the field of mineralogy. But Vernadsky was unsuccessful in his attempts to have Samoilov confirmed as an assistant in his laboratory. Samoilov had been born a Jew; and although he did not practice his religion, considering himself an atheist, he was barred from university employment as a graduate student and from a possible career as a university professor. Only a conversion to Russian Orthodoxy would permit him to work in Vernadsky's lab and prepare himself for a career as a university teacher. Samoilov hesitated to suggest a mock conversion, fearing that a man of Vernadsky's personal honesty and integrity would lose respect for him. For a time, he earned a precarious living by doing chemical analyses for private companies and studying informally with Vernadsky. When he finally suggested a mock conversion, Vernadsky calmed his fears and agreed to act as godfather to Samoilov. Samoilov, in turn, adopted a new patronymic upon his conversion—Vladimirovich, "son of Vladimir," in honor of his godfather. In this way Vernadsky acquired one of his most devoted students. Soon after his "conversion" to the Russian Orthodox religion, to which neither Vernadsky nor Samoilov in reality adhered, Samoilov was confirmed as a graduate assistant at Moscow University without further delay.[150] He went on to become one of Vernadsky's favorites and a productive researcher of those minerals called biolites, products of organic as well as physical processes (coal, peat, etc.).[151] In 1902, Samoilov became the first of Vernadsky's students to receive a professorship in a Russian university when he was confirmed as an adjunct professor at the Novo-Aleksandriiskii Agricultural Institute.[152]

If one gains the impression that an atmosphere of benign paternalism surrounded Vernadsky, that is probably an accurate assessment of the environment he created for his students. However, he personally preferred to see himself as a slightly older and more experienced colleague who worked in a comradely spirit with his students. In reality, Vernadsky was in charge and could crack the whip when he wished, although he generally wielded power in a quiet and indirect way, withholding approval and praise rather than showing anger or disapproval openly or in a manner likely to arouse hostility. And in general, his students seemed to accept and like this form of paternalistic organization. Not only was his lab a focus for group activities, but beginning in the mid-1890s Vernadsky took his students on a series of expeditions every year. Many of these expeditions were to nearby areas, such as rock quarries near Moscow, where he could explain the process of mineral formation in a more natural setting than the laboratory. But other expeditions lasting one to two months and consuming part of each summer were to areas in the Urals, in the Caucasus, and Central Asia.

These expeditions, which sometimes involved dangerous climbs into remote areas, tested the mettle and stamina of both professor and students, and resulted in some important finds. They also created a sense of camaraderie, cooperation, and accomplishment between professor and students. New minerals were discovered, named, and described for the first time to the scientific public. In 1899, one of Vernadsky's expeditions found the first usable deposits of bauxite in the Russian Empire.[153] Vernadsky recognized the importance of the aluminum that could be produced from this deposit. It was in part due to his efforts and those of his student Fersman that the Russian aluminum industry was eventually created. This did not become a reality, however, until after the revolutions of 1917 and civil war, despite their urgings to the Tsarist government before and during World War I.[154]

In 1901, Vernadsky formalized the organization of his scientific school still further by creating a Mineralogical Circle at Moscow University. This group, which included undergraduates, graduate assistants, and young professors, met once or twice a month in the crowded quarters of the Mineralogical Museum to discuss formally the work of individual members of the group. Chairmanship of the group was left open to election by its members, but they insisted on electing Vernadsky, who ran the group in a relatively democratic fashion, reminiscent of the Literary-Scientific Society which Professor Orest Miller had organized at St. Petersburg when Vernadsky was an undergraduate. The Mineralogical Circle, however, was more restricted in scope and membership. The circle had an average membership of around twenty at any given time and provided both an intellectual and social focus for many of its members, particularly given the government's ban on most other kinds of student groups.[155]

In 1903, Vernadsky recognized that his circle of students and former students had begun to form a scientific school, which despite occasional disagreements and differing interpretations, shared a common direction and common goals in mineralogy. In a letter to his wife in April of 1903, he noted that the relationship between himself and his more experienced students, like Samoilov, who were now more independent and had jobs of their own, was changing. The exchange of knowledge and experience was no longer largely in one direction, from professor to pupil. He was beginning to learn more from his students:

> This is a completely new relationship: of pupils working independently, and their *older* comrade. It seems that they, much more than I, consider themselves a school. . . . For me, this is something completely new. To a high degree, much more sharply than before, I feel a sense of responsibility for the character and direction of work—my own and that of my students.[156]

What the Vernadsky school of mineralogy shared in common, which set it apart from most other mineralogists in Russia and the West at the time, was its emphasis on explaining the origins of minerals as part of the earth's history, the evolution of the chemical and physical processes that led to the creation of the mineral kingdom. Unlike the earlier descriptive mineralogy, the new emphasis required of its adherents a solid grounding in chemistry, physics, and mathematics. As one of Fersman's biographers later summed up the history of mineralogy in Russia at the time:

> The end of the nineteenth and the beginning of the twentieth centuries was an especially interesting and significant epoch in the development of mineralogy.

During most of the nineteenth century, mineralogy was a purely descriptive discipline, where a static conception of minerals reigned. The necessary first stage of collecting and systemizing factual material has not yet been completed. In Russia, mineralogy developed chiefly in Petersburg, in the Mining Institute and in the Russian Mineralogical Society which was formed there. The basic direction of the Petersburg school was set by N. I. Koksharov, which made its highest priority the precise study of the physical appearance of minerals. In its time, this direction produced brilliant results, but by the end of the 19th century, it no longer corresponded to the needs of scientific thought, which had progressed beyond it. A huge amount of factual material demanded synthesis and explanation from a new point of view. At this moment the work of Vernadsky played an enormous role. . . .

Preserving the healthy aspects of the old school's traditions in the area of describing minerals, V. I. Vernadsky gave first priority in his research both to explaining their chemical composition and to the problem of genesis, their changes and transformations in various zones of the earth's core. He began to investigate the dynamics of their development.[157]

Mineralogy in general, and the Vernadsky school in particular, was strongly influenced by two crucial developments in nineteenth century European science. First was the growing sophistication of chemistry. This was aided by an arsenal of improved instruments and techniques for chemical analysis which encouraged the development and spread of structural chemistry as well as an increasing knowledge of chemical bonds and chemical reactions, and how molecules form and combine to produce compounds. This was the brand of chemistry Vernadsky had learned in the 1880s from Butlerov and Mendeleev at St. Petersburg University and further perfected in the laboratories of Germany and France. Secondly, Vernadsky and his school were profoundly influenced by the spread of evolutionary views in biology, particularly Darwinism, which had been a matter of intense interest and debate in Russian scientific and intellectual circles from the 1860s on. Evolutionary ideas, transferred to mineralogy, raised important questions about the genesis, development, and interaction among various chemical processes that produce minerals, questions largely ignored prior to the late nineteenth century.

If chemistry and biology were the two leading sciences that influenced the development of the Vernadsky school in mineralogy, physics and particularly the discovery of radioactivity and the debate over the nature of the atom was to have a profound influence on the work of Vernadsky and his students in the early 1900s. Before the discovery of radioactivity, physics had seemed to many a less exciting field in late nineteenth-century science than biology or chemistry, with fewer interesting controversies and a large degree of consensus among physicists regarding the Newtonian concepts which dominated the study of physical forces. The revolution in physics which occurred after the mid-1890s produced a new direction in the Vernadsky school which resulted in the creation of several new fields in Russian science where Vernadsky and his associates played a key role, particularly geochemistry, biogeochemistry, and radiology. But before we examine these developments, we must turn to another revolution which intervened and in which Vernadsky himself played a significant role: the Russian Revolution of 1905 and its aftermath which were to shape and influence the development of Russian science in important ways.

CHAPTER

3

THE POLITICS OF MORAL INDIGNATION
Vernadsky in Science and Politics, 1903–1914

On April 23, 1903, Vernadsky wrote to the Golshteins in Paris: "The terrible and depressing years of university life take much of my time and force me to spend my strength in university service. This year things have slackened off a bit. It is difficult to say what the near future will bring."[1] As it turned out, the spring of 1903 was a brief calm before the stormy events of the next decade, which placed their stamp so strongly on Russian society and deeply affected the development of Russian science and higher education. In particular, these events were to widen the rift between most of the scientific community, which was liberal or radical in its politics, and the Tsarist government. In the absence of much private support for science, this made it even more difficult for academic scientists (those in higher education and the Academy of Sciences who were the large majority of professional scientists) to work together with the government. Suspicions and mistrust, already at a high level by 1903, hardened over the next decade.

Although Vernadsky managed to remain active as a scientist, maintaining his lab at Moscow University through the turmoil of the Revolution of 1905 and keeping together his circle of students and younger colleagues during these difficult times, his rate of scientific publishing understandably diminished in the years between 1904 and 1907. Vernadsky's energies were increasingly drawn into politics. Instead of scientific articles, much of the product of his pen in these years consisted of political editorials, newspaper articles, and pamphlets on public issues. Most of his writings were imbued with a deep sense of moral indignation at the condition of Russian society and the conduct of the Tsarist regime, which Vernadsky and fellow members of the intelligentsia held responsible for Russia's backwardness. In the process, Vernadsky proved to be a master of the politics of mobilizing moral indignation among his colleagues. His mastery of this "art" was one of the major factors, albeit not the only one, that helped to bind him and his friends and associates together into an enduring network that lasted for many decades. During these years, Vernadsky himself was transformed from a local activist of mildly liberal persuasion and limited reputation to a national political figure who stood among the inner circle of those who forced major concessions from the autocracy in the fall of 1905 and then watched angrily and for the most part helplessly as the Tsarist government attempted to limit those concessions in the years after 1906.

Although less dedicated to politics than some others in the inner circle of the liberal movement (such as Struve, Miliukov, and his close friends Shakhovskoi, Kornilov, Rodichev, and Petrunkevich) and less flamboyant as an orator and public figure, Vernadsky nonetheless was a thoughtful and active member of the Russian constitutionalist movement with influential opinions about university self-rule, public education in general, agrarian land reform and capital punishment. He expressed these opinions not only in the inner circle of the liberal constitutionalist movement but also in the press and the State Council, the upper house of the new Russian parliament, to which Vernadsky was elected between 1906 and 1911 as one of a half dozen representatives of the universities and Academy of Sciences.

Vernadsky was one of twenty men who founded the Union of Liberation in July 1903 (a coalition of liberals and radicals dedicated to a public campaign aimed at peacefully ending the autocratic regime in Russia.) His apartment in Moscow became a center of this illegal and conspiratorial movement during the next several years. Vernadsky went on to become a founder and member of the Central Committee of the Constitutional Democratic Party in October 1905, and he remained an active member until 1918. He also became one of the principal organizers of the Academic Union, which, at its height in these years, enrolled more than half of all the university teachers in the country in a professional union aimed at university self-rule.

Vernadsky's political activity in these years was in a sense the logical culmination of those values he had professed while a university student twenty years earlier and those personal friendships he had formed then in the Oldenburg circle. In fact, one might argue that he was drawn so heavily into politics in the early years of the twentieth century primarily at the urging of his closest friends, such as the Oldenburg brothers, Dmitrii Shakhovskoi, A. A. Kornilov and their associates in the zemstvo constitutionalist movement like Fedor Rodichev and I. I. Petrunkevich, who dragged him reluctantly from his preoccupation with scientific work into political activity. Certainly there would be some truth to such an assertion, judging from a number of statements in his correspondence from these years in which he frequently expressed his preference for scientific work over social and political activities.[2] However, to stop with such an assertion would be to miss a deeper underlying affinity between Vernadsky's commitment to science and his political activism. While in one sense a distraction from his scientific work, his political commitment, in another sense, was strongly related to his scientific world view. This becomes clearer when one compares some of his writings on the philosophy of science with some of his political journalism from this period.

In 1902–3 Vernadsky wrote and published a book entitled *On a Scientific World View*.[3] Based on a course he gave at Moscow University which was intended to familiarize his students with general ideas in the history and philosophy of science, the book contains a view of scientific knowledge and the process leading to it which is closely related to his view of politics. In that book Vernadsky developed a concept of scientific truth which was derived from and very familiar to liberals in the West: all scientific knowledge is partial and tentative and must be subject to the competition of ideas and the test of criticism. Conflict and controversy in science, therefore, is beneficial because it leads to new knowledge that may come closer to explaining the existing facts and producing a "truer" understanding of nature. Hence, we need tolerance and moderation in the search for truth, since we have frequently been under the influence of false ideas in the past and may continue to be so even today. From this, he concluded that we should be skeptical about any institutions or individuals who claim exclusive truth and seek a monopoly for their views.[4]

In the same way, Vernadsky's view of politics was closely related to his view of scientific truth. In an article published in 1906, Vernadsky condemned the terrorism of both the right and the left in Russian society, since both ends of the political spectrum were attempting to use force to make their views dominant. Vernadsky called for tolerance and a peaceful competition of political parties within a representative system. In his view, such an open clash of goals and viewpoints could lead to the working out of the best solutions for Russia's future. He put his faith in the initiative of different social groups, who if permitted to organize freely, would protect and promote their own interests and, through a process of competition and mutual compromise with other groups, achieve realistic solutions for his nation's problems. "The more energetic the organization of parties, the wider they encompass Russian life—the sooner and more completely will the anarchy end," he wrote in 1906.[5]

Vernadsky's view of representative democracy may seem idealistic and even naive to a later generation more aware of the imperfections of such institutions. His naiveté, however, was in fact tempered by many years of practical experience with three types of representative institutions in Tsarist Russia—first as an active elected member from 1890 until 1913 of local self-government in Tambov province, then as a member of the faculty council of Moscow University during much of the same period, and finally as an elected member of the upper house of the Russian legislature, the State Council, from 1906 to 1911 as a representative of the universities and the Imperial Academy of Sciences. Despite occasionally intense frustration and ideological setbacks, he remained—true to his generation and circle of friends from university days—idealistic and optimistic. His idealism and optimism in politics was in turn strongly influenced by his scientific worldview.

Vernadsky's worldview developed not only in the Oldenburg circle and in his work as a scientist, but also in his conversations with Lev Tolstoi in the early 1890s. As he noted in a diary entry for 1893, "I think that the teaching of Tolstoy is much deeper than I first thought. This depth consists (in his view of) 1. the basis of life is the search for truth; 2. the real task is to tell this truth without any retreats. I think that the last is very important and that the denial of any hypocrisy and pharisaism forms the basic strength of the teaching, since as a result the personality becomes stronger and the personality receives social force."[6] Vernadsky came to see Tolstoi as a major accumulator of the "energy" of human consciousness and reason, and it was the progress of consciousness and reason, through science and higher culture in general, that Vernadsky by the early 1890s had come to see as his personal goal. This led him inexorably to invest his energies in scientific research and secular education and struggle for a political system that would further these goals. Vernadsky did not have a naive belief in the inevitable progress of reason and consciousness, although he did believe that such progress had been the general trend of the nineteenth century. Rather, he believed that life consists of a constant struggle between regressive and progressive elements. Uncertainty about the outcome of this struggle compels a personal commitment from those who wish to be on the side of human progress:

> In essence, we see in all of history a constant struggle between the conscious structure of life against the unconscious structure of the lifeless laws of nature—and in this effort of consciousness rests the beauty of historical phenomena, their originality among the other natural processes.[7]

At another point in the same period, he wrote:

> I do not understand life without consciousness. It seems to me like this: there are in life regressive currents which lead men to a heavy, sensual, animal life. These currents put their stamp on everything which happens. . . . But in this same life there is another element, a small kernel of progress—the force of consciousness. It pushes upwards, it is the result of a huge quantity of work expended on the earth for working out higher forms. . . . There are people who carry the banner of consciousness; and the life around it has been deeply penetrated by them . . . woe to that country where such people hide the fire [of reason] which burns in them and distort its sacred activity. The country where people can develop their consciousness will be stronger.[8]

This view was closely tied to Vernadsky's nationalism and his criticism of the Tsarist government for the obstacles it created for Russia to realize its potential strength and creativity. His emphasis on the role of creative individuals in advancing knowledge might have led Vernadsky, as it led some thinkers of his time, to believe in an aristocracy of the spirit, superior to and above the "heavy, sensual, animal" existence of the masses. Indeed, at times one catches a whiff of *noblesse oblige* in Vernadsky's writings and actions. But given the risks he took and the great dedication of time and effort to promoting mass education, democratic reforms, and a more egalitarian distribution of property, there is no reason to doubt the sincerity of his democratic beliefs. In fact, Vernadsky struggled to reconcile his belief in the role of individual geniuses who advance mankind's rationality and control over the environment with the role of the masses.

At times, he saw the masses themselves as carriers of the consciousness-transforming principle of social progress; it did not come from some kind of aristocratic spirit set above them. At the same time, he believed that separate outstanding personalities can express this mass consciousness with special force and consequences. Such individuals can take on themselves the responsibility of developing spheres of consciousness in science, art, and philosophy.[9] What his view boiled down to was a belief that individual geniuses, pioneers of cultural advancement, ultimately are dependent on the masses; their discoveries can only be applied and sustained if the masses themselves are sufficiently supportive and aware of the benefits to be derived from such innovators. Thus there is a mutual interaction between the masses and their cultural benefactors. While the masses provide support for the innovators, the innovators and the educated in general have a debt to repay to the masses by working to raise the general cultural level so that the population at large can absorb and perpetuate these advances. This idea, ultimately derived from populist thinkers of the 1860s like Lavrov, was embedded in Vernadsky's thought. As he expressed it to his wife in 1893: "I am deeply convinced and become ever more convinced that the sole possibility of making culture durable is to raise the level of the masses, to make culture a necessity for them."[10] A year later he wrote to his wife again along the same lines: "Woe to that country where knowledge is poorly developed, where it has barely penetrated the working masses."[11]

Ten years after writing these words, Vernadsky took what he considered a further decisive step toward the realization of his ideal, becoming one of the twenty founding members of a conspiratorial organization, the Union of Liberation. Not a mass organization but a popular front of change-oriented intellectuals, the founders were mostly friends and longstanding acquaintances who had worked together in the universities, in famine-relief, literacy circles, and local self-government. These twenty men met in July 1903 on the shores of Lake Constance in Switzerland, away from the prying eyes of the Tsarist secret police, and agreed to launch a peaceful and public campaign to transform the Russian autocracy into a more democratic and representative political system. As one historian of that movement has recently written, "an organization was created which had as its express purpose to lay seige to the autocracy by means of a massive and overt public campaign. A coalition of liberals and radicals, it was the only successful revolutionary organization in the history of Imperial Russia. Before two years passed, it succeeded where all the others had failed: it forced the autocratic regime to give up its monopoly on political power."[12]

What this assessment leaves out, of course, is the great personal risk these men took in forming such a political organization and their good fortune that such an organization came into being at an opportune moment. In July 1903 they could not foresee that the Tsarist regime would soon become embroiled in a highly unsuccessful war with Japan which would destabilize the autocracy and allow the Union of Liberation to provide leadership and direction for a mass movement that the government, temporarily at least, would be too weak to resist. This movement culminated in a partial sharing of power with representative institutions on a national level and a partial liberalization in the area of civil rights and university autonomy. These concessions were eventually retracted as much as possible by the regime as it regained its power and confidence after 1906. Such qualifications notwithstanding, the accomplishment of the Union of Liberation was considerable when viewed in the light of earlier reform and revolutionary movements in Imperial Russia. It is therefore of considerable interest to learn what we can of the motivations and actions of its core leadership.

In Vernadsky's case, it is revealing to know that in the same month he helped to found the Union of Liberation, he was writing an introduction to one of his lectures in the series, "On a Scientific World View," entitled "The Progress of Science and the Popular Masses." Although not himself a socialist, he saw an alliance developing between working class movements and science, with scientific thinking providing the intellectual tools for those who believed that society could be refashioned in a more rational way. Vernadsky considered socialism the first mass movement in the history of Europe and America to be self-consciously under the influence of modern science. Although Vernadsky was not familiar with Marxist writings during this period, he found much that was true among the writings of the first socialist thinkers, particularly the utopian socialists like Owen and the Saint-Simonians.[13] While remaining outside the socialist camp, Vernadsky and his associates in the Union of Liberation clearly saw no serious barriers to a popular alliance between liberals and socialists. In fact at this time, they formulated the general principle of the liberation movement—"no enemies on the left."[14]

For Vernadsky at least, this alliance was strengthened by his belief that the development of science in modern society depended upon the process of political democratization. As he indicated to his wife in July 1903, while writing "The Progress of Science and the Popular Masses," "I consider that the interests of scientific progress are closely and inextricably tied to the growth of a wide democracy and humanitarian attitudes—and vice versa."[15] Vernadsky saw a mutual interdependence between democracy, humanitarianism, and scientific progress. His belief in democratization as an impetus to science was more an article of faith and an assertion scattered in his writings than a carefully reasoned argument; and Vernadsky left no comment on the counterargument that science might flourish better in an elitist society dominated by men of reason, such as a Platonic utopia or the technocratic utopias developed by socal thinkers like Saint-Simon in the early nineteenth century and Thorstein Veblen and his successors after 1919.

While not supported by a closely reasoned argument in his writings, this positive faith in the links between democracy and science sustained Vernadsky in these difficult years of conflict with the Tsarist regime. His belief provided a conscious rationalization for his hatred of the autocracy and helped to justify his struggle against the Tsarist system of rule by courtiers and what he and his friends considered a largely ignorant bureaucracy, accountable neither to the educated elite of Russian society nor to the masses. Whether a democratic system of government would, in fact, support science and in turn be supported by science was as yet unproven. What Vernadsky and his associates knew from their own long and bitter experience was that the work of scientific research and education was impeded, indeed made extremely difficult at times by conditions which they blamed on the Tsarist system.

This opinion was strongly reinforced by work experience in the two institutions centrally concerned with the progress of science and education in Tsarist Russia: institutions of the zemstvos and the universities. Most of the leaders of the Russian constitutionalist movement in the early twentieth century had been active in one or both of these institutions for more than a decade. Vernadsky was one of those who was active in both and saw a strong professional and cultural link between these two institutions. By 1900, the zemstvos employed more than 50,000 graduates of Russian higher education, the so-called "third element" (physicians, statisticians, agronomists, teachers, engineers, etc.). These 50,000 specialists represented more than half of all persons with higher education employed in the Russian economy at the time.[16] Zemstvo liberals like Vernadsky considered such professional employees their natural allies against the central bureaucracy and against the fiscally conservative rural gentry, who were primarily interested in using the zemstvos to improve agriculture on their estates but opposed increasing zemstvos budgets in order to expand social services for the general population.[17]

Vernadsky became involved in the zemstvos by virtue of ownership of his estate. Vernadovka was located along the main rail line between Moscow and Kazan and could be reached in a matter of hours by passenger train from Moscow. Vernadsky had inherited this estate from his father, who had in turn inherited it from his first wife, Maria Shigaeva. During the 1890s, most of the lands of this estate were rented out to local peasants, who raised grain and hay and paid Vernadsky rent. An estate of this size was considerably larger than that held by the average nobleman of this region and, of course, many times larger than most peasant holdings. A middling peasant might hold forty acres, and a well-off peasant one hundred twenty-five acres.[18] Yet Vernadsky's estate, huge by comparison with average peasant landholdings, was itself dwarfed in size by those of the upper nobility such as the Manuilovs, who held over four thousand acres in the same county, Count Mordvinov, with over 50,000 acres, and Countess Shuvalova, with more than 45,000 acres.[19]

Given the falling grain prices in the 1890s, the income from his estate in this period was modest, and Vernadsky reinvested most of it in the property, with the remainder used to help pay for his summer visits to Western Europe, books, and help to friends and relatives.[20] With his future in the Russian university system a nagging uncertainty, Vernadsky spent much of the decade from 1890 to 1900 improving Vernadovka, including the construction of a house for his family to move to permanently should that become a political necessity.[21] Vernadsky viewed this land as an insurance policy against possible dismissal from the university. Beyond that, as he commented to his wife in 1896, it had become difficult to sell land in Tambov, given falling grain prices, periodic famine, and unrest.[22]

Vernadsky was first elected to the local Morshansk zemstvo in 1892, following his active role there in helping to organize famine relief. He served in that body until 1913, except for a three year hiatus, 1907–10, at the height of the reaction by conservative noble landowners against liberals like Vernadsky.[23] He was also periodically elected a member of the Tambov provincial zemstvo. Although he lived in Moscow during most of these years, he conscientiously commuted by rail to attend zemstvo meetings and took an active part in many zemstvo commissions concerned with budget, public health, primary education, and agriculture.[24] He also had a number of clashes with the authorities and with conservative noblemen as a result of his zemstvo activity in this period. As he wrote to his wife in 1892, he was the only member of the Morshansk zemstvo to speak out against closing down local primary schools as a result of a deficit in the budget, a deficit brought about by the famine and generally unfavorable economic conditions.[25] When the head of the Russian Orthodox Church, Pobedonostsev, (who was Procurator of the Holy Synod and one of the Tsar's closest advisors), opposed the opening of secular Sunday schools by the zemstvos, Vernadsky sued him in the highest court of the land, the Senate, and won.[26] He eventually built such a school on his own land and gave it to the local zemstvo as an example to other local landowners.[27]

Throughout the decade of the 1890s there were regular skirmishes with the bureaucracy and with the more conservative local nobility. For example, when he was elected a local justice of the peace in 1892, the Ministry of the Interior failed to confirm his appointment. Nonetheless, at the beginning of his zemstvo service, Vernadsky was frequently in an upbeat mood: "One can learn a great deal participating in zemstvo meetings; I never imagined what a useful and important school this is for everyone," he wrote in 1892.[28] Yet even Vernadsky was worried that the occasional liberal moods of the zemstvos were often not matched by a commitment to provide adequate financial resources. As he wrote again in 1892 to a friend, "I have just returned from a provincial zemstvo meeting. It produced a strange impression on me. The meeting was undoubtedly liberal but at the same time it did not represent the interests of the whole population but of only some of the tax payers; and in general it was evident that they were everywhere concerned with the defense of their pocketbooks."[29]

Despite his pride in the opening of many new primary schools by the Morshansk zemstvo and the establishment of a new zemstvo hospital, by 1900 Vernadsky was writing his wife that he found it harder and harder to work in the zemstvos.[30] Part of the problem was simply the strain of combining such volunteer work in an outlying province with a busy career at Moscow University. Vernadsky felt that he had spread himself too thinly over the past ten years and was guilty of dilettantism.[31] But much of the problem was the result of the petty harassments imposed on zemstvo liberals by the central government.[32]

Despite the frequent feeling of pushing a boulder uphill, Vernadsky did not give up and was even elected secretary of the Morshansk county zemstvo in 1900.[33] In 1899, he helped establish a hospital in the county and in 1904 he proposed to the Tambov provincial zemstvo the establishment of a weekly zemstvo newspaper, a proposal which was passed by a large majority.[34] But perhaps the most significant event in Vernadsky's zemstvo career came in May 1902, when he became one of twelve persons elected by zemstvo activists to arrange regular national conferences of the zemstvos, an action that had been expressly forbidden by the Tsarist government for decades. This group of twelve included other zemstvo liberals like Shakhovskoi and Petrunkevich, who had long been friends of Vernadsky. It also led directly to a confrontation between the zemstvos and the central government. Zemstvo liberals became less afraid to defy the autocracy openly and call for its transformation into a constitutional and parliamentary democracy. The national zemstvo movement helped to launch Vernadsky's career as a national political figure, active over the next several years in organizing the Union of Zemstvo Constitutionalists, the Union of Liberation, and the Academic Union. In October 1905, the creation of a political party, the Constitutional Democratic party (Kadets), resulted as a logical outgrowth of these organizations.

As must be evident by now, Vernadsky was part of a cohort of men of similar age and background who had risen to leadership positions in the zemstvo movement between 1892 and 1902. These were "men of the eighties," with university educations, often in the natural sciences, and members of the no-bility by social origin but liberal or radical by conviction. An early historian of the zemstvos, Veselovsky, identified by name some 120 "men of the eighties" who became active in the zemstvos in this period, the majority of whom became members of the Kadet party.[35] Ironically these men owed their rise to promi-nence in the zemstvos, at least partially, to one of the counterreforms of Alexander III, which had been aimed at stifling liberal and radical currents. The Zemstvo Act of 1890 had replaced the zemstvo electoral system, which until then had been based on landed property, not social origin, with a system based on election by social estates.[36] This act gave the predominant political role in the zemstvos to about twenty thousand electors from the hereditary nobility.[37]

The intention of this counterreform had been to diminish the role of peasants and non-noble landowners in the zemstvo and increase the role of the conservative rural gentry. Whereas it did give the latter element a firmer veto than before over the actions of the zemstvos, most of the conservative rural gentry tended to be apathetic and avoided zemstvo service. This opened the way for a small but active group of liberal and radical hereditary noblemen, like Vernadsky and his friends, to increase their influence. They were willing to invest considerable time in these institutions, motivated more by ideological than material interests. They shared leadership of the zemstvos with another group of noblemen, who tended to be more moderate, but still shared much of the disdain of the liberals and radicals for the Tsarist bureaucracy. These moderates were part of a movement among the younger nobility who sought a role for themselves not as officials but as "improving landlords," who deliberately chose to live in the countryside and farm their own estates rather than renting their land to the peasantry and living in one of the large cities as absentee owners.

This latter movement was strengthened by the fact that after 1896 the long depression in grain prices ended, and by 1905 grain prices had risen some 48 percent over prices in the mid-1890s.[38] Although many of the nobility sold off their lands between the abolition of sefdom in 1861 and 1917, there was a countertrend between the 1890s and 1917. By 1917, according to Manning, 68 percent of arable gentry land was being cultivated by the gentry as opposed to only 33 to 47 percent at the end of the nineteenth century.[39] The interests of this group of moderate "improving landlords" were strongly economic, and, therefore, potentially in conflict with a land-hungry peasantry, especially in overpopulated blacksoil regions such as Tambov province, where Vernadsky's estate was located.

It was in part due to the influence of this group of noblemen that the zemstvos increased their spending on agricultural services by some 400 percent between 1894 and 1904. These "moderates" also shared certain interests with liberals like Vernadsky and his friends: for example, a dislike of the Tsarist bureaucracy and a belief in some form of representative institutions which would increase their influence on the central government and help to counter the bureaucracy. Thus, moderates tended to support the zemstvos and the creation of some form of national representative institution like a parliament or consultative assembly, so long as such institutions were dominated by the nobility and acted in their interests. In this respect, they were more narrowly class-oriented than people like Vernadsky.

Liberals like Vernadsky continually tried to reach out to the "third element" of professional employees in the zemstvos and to non-nobles, including the peasantry, for support.[40] They sought to extend the social services of the zemstvos to more of the general population. The "moderates" tended to look inward, toward their own class of rural gentry and toward protection of their class privileges. They were more fiscally conservative and supported services that directly benefitted their group economically, such as agricultural services.

Most of the "moderates" eventually ended up as members of parties further to the right of the Kadets after 1905, such as the Octobrists and the Russian Nationalist party, allied with Stolypin after 1907.

Nonetheless, until 1905, there was a partial alliance between liberals and moderates in the zemstvo movement. The liberals, in fact, became increasingly dominant as leaders of the zemstvos between 1902 and 1905 as the confrontation with the autocracy developed in these years. Many members of the rural gentry, including the moderates, were angered by the minister of the interior's prohibition on the discussion of non-local issues by the zemstvos in 1903 and again in 1905. As a result, they often supported liberals, who swept the zemstvo elections of 1903–1904 and increasingly used the zemstvos as a forum to agitate for a constitutional and parliamentary government to replace the autocracy. By the winter of 1904–5, four-fifths of all the zemstvos and two-thirds of all the provincial nobles' organizations supported some form of representative government in Russia.[41] There was, however, a split which foreshadowed later political conflicts. The split involved those nobles who wanted a parliament with real power and those who still supported the autocracy but wanted an elected national assembly, dominated by the nobility, which would have consultative powers to advise the Tsar.

The Union of Liberation, dominated by the friends and associates of Vernadsky, took over preparations for the 1904 national zemstvo congress, which decided by a vote of 71 to 27 to work for the immediate establishment of a constitutional regime in Russia. This congress appealed to local zemstvos for support in an address campaign, and large crowds of non-zemstvo men attended these local assemblies, temporarily intimidating more moderate and conservative noblemen.[42] The increase in peasant violence in 1905 drove a wedge between the majority of rural gentry, who were dependent on farming and their estates for a living, and the liberal element represented by Vernadsky, who tended to be urban-based professionals. Peasant unrest was often directed against the property of "improving" landlords who farmed their own estates. Against the background of rural upheaval, the political rift between the rural gentry and the liberals escalated due to the fact that liberals by and large supported forced expropriation of gentry lands (with compensation paid by the state), in order to satisfy the land hunger of the peasantry.[43]

These events drove more and more of the rural gentry, including the moderates, back into the arms of the autocracy, the traditional guarantor of their privileges and economic interests. This was especially true after the October Manifesto of 1905 promised some form of representative government and thereby (they hoped) greater influence by the rural gentry on national policies. By 1906–7, the moderates and conservatives among the landed nobility had turned against liberals like Vernadsky and his friends in the Kadet party.[44] The zemstvo elections of 1906–7 swept most of the liberals, including Vernadsky, from office in the zemstvos, which now became a bastion of class conservatism, protecting the interests of the rural gentry from the peasants and liberal reformers whose program threatened the economic interests of the "toiling nobility," as these farmers among the gentry liked to style themselves. [45]

If 1902 marked an upsurge in activism among zemstvo workers like Vernadsky, it was in 1901–2 that university professors began to awaken from their Rip van Winkle slumber of the previous several decades. While he was making a name for himself with his zemstvo leadership, Vernadsky gained national prominence as a leader in higher education. The previous five or six years had seen increasing student disorders that were caused by the tight regime of surveillance imposed by the government on institutions of higher education and student resentment of this regime, which resulted in frequent clashes between police and students. By 1899, this led to the largest student protest ever to occur in the Russian Empire. Some 30,000 students in higher education went on strike throughout the Empire, the large majority of all such students. Thousands were expelled from the universities for participating in disorders, and on July 29, 1899, Nicholas II approved the so-called "temporary rules" that drafted expelled students into the army. [46]

In March 1899, Vernadsky wrote in his diary about these disorders:

> All this time I have wanted to write in more detail about our affairs here in Moscow. But the pen drops from my hand; it is all so depressing and I feel completely helpless. Our university is closed; all the entrances are locked; everywhere the doors are guarded by custodians, the university police, and today real policemen at two gates. Almost all the students submitted apologies, but these were not accepted: they say that 809 are suspended until next fall. These are days of exile from Moscow; and every morning the trains are overflowing with expelled students. All this time the professors stand completely on the sidelines; the Faculty Council has not met a single time; once at the beginning of the disorders the professors were called together for a private conference with the rector . . . All this is very depressing because power now is in the hands of people who are completely removed from and foreign to the interests of young people; and they are acting quietly in the most severe manner. . . . In general the apathy of "oppositional" forces is amazing—no one acts or wants to act. [47]

Student passivity proved temporary, however, and within two years the situation had begun to change. The academic year 1899–1900 passed relatively quietly, with many students doing military service after having been expelled from the university. But on February 14, 1901, an expelled student shot and killed the minister of education, Bogolepov. Bogolepov's assassination reactivated student radicals who, a few days later, organized a demonstration in front of the Kazan Cathedral in St. Petersburg to mark the anniversary of the emancipation of the serfs. When mounted police appeared and whipped the 400 demonstrators, sympathy strikes and demonstrations began to spread to higher educational institutions in other parts of the Russian Empire. [48] Professors at Moscow University appealed to their students not to strike, since they argued that such strikes were counterproductive. The curator of the Moscow Educational District, Nekrasov, "realizing that the professors could be used as a moderating influence, on his own authority agreed to waive the strictures of the 1884 University Charter and allow the Faculty Council to appoint a special commission to investigate the causes of student unrest and make appropriate recommendations for reform."[49] Nekrasov's initiative proved to be a wise move, since it averted a general student strike. In addition, the Faculty Council succeeded in its appeal that students arrested in that winter's disorders not be drafted into the army. Later in the year this faculty commission "issued a bulky report which recommended that the inspectorate's power be curbed, that student organizations be legalized, and that the Charter of 1884 be reconsidered."[50] Vernadsky, who was not appointed to this faculty commission, probably because the curator of the Moscow Educational District considered him politically unreliable, nonetheless submitted two memoranda to the commission, one in his own name and one which he signed with nine other distinguished professors in the natural sciences, including Timiriazev, Sechenov, and Umov.[51] They protested the increase in police repression and government surveillance of university life.[52]

In 1901, Vernadsky took a further step to involve himself in the question of university reform. He published a pamphlet in which he recommended the restoration of the university charter of 1863 with its guarantees of autonomy for the university faculty and their right to elect their own deans and rectors.[53] During 1901 and 1902, Vernadsky began to play a much more active part in the Faculty Council. Hopes for educational reform were raised by the actions of General P. S. Vannovsky, who was appointed in mid-March 1901 to succeed the assassinated Bogolepov as minister of education. A number of conciliatory gestures toward students and faculty accompanied the appointment of Vannovsky. In April, the minister promised a thorough review of the charter of 1884 and sent a list of questions to faculty councils throughout the Empire, soliciting their opinions.

Vernadsky wrote to his wife in May, expressing cautious optimism about this latest development. He felt it was hard to say if anything would come of it, especially since responding to the government's overtures would require good people in the university to carry through; and for several years previously Vernadsky had expressed his criticism of faculty passivity.[54] Two days later, Vernadsky mentioned again that the Vannovsky circular was causing excitement among the faculty; and on May 8 he wrote again, indicating that the Faculty Council had begun to discuss basic university reforms as a result of Vannovsky's questionnaire. However, Vernadsky expressed his disappointment that the most articulate faculty members left the faculty meeting early, although he felt compelled to speak out himself.[55] By May 13, Vernadsky reported that the Faculty Council had taken a definite stand in favor of replacing university administrators who were appointed from above with rectors and deans elected by their faculties.[56] By the summer of 1901, Vernadsky was deeply involved in such questions, agreeing to prepare drafts of written answers to several of the questions in the Vannovsky circular, specifically the questions concerning required government exams in subject-matter areas, the legal rights of persons completing the university, and the question of how to bring students and professors closer together.[57]

Vernadsky's early caution about Vannovsky's overtures proved to be justified. The elderly general turned out to be less liberal than students and faculty had hoped, yet too liberal for Nicholas II and some of his top bureaucrats, who had been looking to Vannovsky for a "quick fix" of the universities, that is, a rapid return to order on the campuses. Nothing came of Vannovsky's early hints of university reform; and by April of 1902, he had been removed from office by the Tsar, partly in response to the assassination of the minister of the interior, D. S. Sipiagin, and partly due to the Tsar's opposition to Vannovsky's plan for reform of secondary education.[58] The one concrete change introduced by the government into the universities during Vannovsky's tenure were the temporary rules on student organizations, adopted in November 1901. These rules permitted the creation of such student organizations as discussion groups, clubs, credit unions, reading rooms, etc. but only under the direct supervision of a faculty member appointed by the rector. The rector could intervene to dissolve such organizations or to interfere in their functioning if he chose. General student meetings were prohibited, although an earlier draft of these rules had legalized skhodki conducted under close university supervision. The minister of interior, Sipiagin, opposed such legalization and the provision was deleted.

Although the temporary rules on student organizations permitted many activities which had earlier been strictly illegal, students were bitterly disappointed at their restrictiveness. Faculty were generally not pleased by their new role as policemen for these organizations, and the Council of Ministers was itself divided over the possible effects of the rules. At its meeting of November 18, 1901, Vannovsky hoped that the rules "might pacify the students, but added that he had grave doubts that this would in fact occur. Witte was convinced that the temporary rules would only lead to new disorders, but the Government had little choice but to publish them anyway. Only Sipiagin managed to retain some degree of optimism. The rules *would* touch off new disorders, he admitted, but that would enable the government to arrest the radical ringleaders more easily."[59]

Witte and Sipiagin proved correct. The hopes of moderate students for a less repressive reform were dashed, and the hand of more radical students was strengthened. The winter of 1902 was marked by frequent student disturbances protesting these rules, disturbances which were crushed by widespread arrests ordered by Sipiagin and carried out by the Okhrana. In the process, Sipiagin was assassinated by a militant wing of the Socialist Revolutionary party, and Vannovsky was abruptly fired by the Tsar, to be replaced by a high level bureaucrat in the Ministry of Education, Zenger, who himself lasted less than two years. By 1904, it was clear that the government lacked a consistent policy in higher education. The government's actions had only raised hopes for change among both students and faculty and then frustrated those hopes with a policy of renewed repression. Repression worked temporarily during the academic year 1902–3 to bring calm to university campuses, but at the cost of alienation among reform-minded faculty and smoldering resentment on the part of a large number of students, among whom the radicals had gained considerable influence.[60]

One of the most perceptive studies of Russian universities in this period concludes: "In the fall of 1901, the government had a chance to pacify the moderate majority of students, it did not do so. The Ministry of the Interior was able to force the acceptance of a policy based on repression and only half-hearted concessions. Both in its policy toward the professoriate and in its policy towards the students, the Government between 1899 and 1902, pursued a confused and contradictory course. The crisis of the universities was a long way from being over. The academic year 1901–2 was a turning point in the history of Russian student protest. Henceforth, the majority of students would no longer trust so readily in the good intentions of the government."[61] The same could be said of liberal professors like Vernadsky. If they had been reluctantly willing to give the government the benefit of the doubt for a short time in 1901, their experience over the next several years reconfirmed their suspicions of the government's intentions toward the universities.

One of the most serious bones of contention between the government and professors concerned the question of student discipline. In August 1902, the Ministry of Education had decreed the establishment of disciplinary courts for students, composed of professors elected by the faculty. Making professors responsible for student discipline had been one of the provisions of the university charter of 1863 which had been abolished in 1884. Although many professors were ambivalent about once again accepting this responsibility, they nonetheless took over the duty as a step toward university autonomy, with the assumption that the central government would not intervene. Their assumption was mistaken. When the disciplinary courts proved too lenient toward students involved in disturbances, too lenient at least from the point of view of the ministry, the central government intervened, for example at Odessa, Warsaw, and Kharkov universities in 1904–5. Such interference angered both students and faculty, as did the measures taken in early 1904 by the new minister of the interior, von Plehve, to end the vestiges of autonomy and special privileges enjoyed by the higher technical institutes which were under his jurisdiction.[62] Von Plehve's actions led to student disturbances at these institutions and their temporary closing in the spring of 1904.

However, for much of 1904 the anger of students and faculty simmered but did not boil out of control, in part because of an upsurge of patriotism in Russian higher education brought about by the Japanese attack on Russian possessions in the Far East in late January 1904. This upsurge of patriotism lasted until the fall of 1904. Just prior to the Japanese attack, the Union of Liberation held a conference in early January 1904, attended by Vernadsky and other liberals, which declared as its aim the replacement of the autocracy by a constitutional regime.[63] But a few weeks later, after the Japanese attack, groups like the Union of Zemstvo Constitutionalists, closely affiliated with the Union of Liberation, declared at its congress in February 1904: "Because of the upsurge of patriotic feelings in the country it would be impractical to put forward constitutional demands or to ask the zemstvo assemblies to incorporate such demands in their petitions."[64] Soon after the Japanese surprise attack, the faculty councils of all nine universities sent resolutions pledging their loyalty to the Tsar in this war.[65]

Soon after the start of the war with Japan, Vernadsky wrote his colleague Sergei Trubetskoi, professor of philosophy at Moscow University and soon to be the first freely elected rector of that university in many decades, complaining about those who hoped for defeat of their nation by Japan. Vernadsky made a distinction which had already become important in his mind between the Russian state and the existing autocracy. He wished to build up the power of the state while seeking reforms of the existing government. No state could survive, he indicated, "unless it was able to use science and technology in a rational and efficient manner."[66] In expressing this atittude, there is no reason to believe that Vernadsky was at odds with the majority of the liberal professoriate, who favored a strong state, one run by competent, well-educated people. At any rate, the level of political activity aimed at establishment of a parliamentary government diminished considerably during the first half of 1904, as such liberals rallied behind the Tsarist government in time of war.

Unfortunately, the Tsar did not give the liberal professoriate the same benefit of the doubt. In April 1904, Nicholas II appointed another elderly general as acting minister of education, V. G. Glazov. At that time director of the Army's General Staff Academy, Glazov had no civilian experience with education, having spent nearly four decades in the military. When told he might be appointed to the Ministry of Education, he confided that he knew nothing about the ministry and had "no particular educational policies in mind."[67] In his audience with Glazov in April, Nicholas II, who had dismissed two ministers of education over the previous two years, declared that "right now the Ministry of Education is in such a state of chaos that we must undertake most energetic measures . . . first of all we have to clean house from the top, beginning with the professors: there are decent people in the professoriate, but very few."[68]

Glazov proved to be no improvement over the previous two ministers. In September 1904, he recommended to the Tsar a limitation on the disciplinary role of faculty courts in the case of mass student disturbances. In such cases, rectors, in consultation with the curators of local educational districts, could bypass faculty-controlled courts entirely and could expel or suspend students on their own authority, without seeking faculty approval. The Tsar agreed with Glazov and adopted this new policy, which much of the professoriate considered a direct attack on their competence and integrity.[69] Yet only nine months later, in June 1905, Glazov came to consider the new policy a failure and reversed it, returning student discipline to the control of faculty courts with the hope that by doing so "future student disturbances would be interpreted as being directed against the professoriate rather than against the government."[70]

Even before this latest reversal in policy, a new phase was inaugurated in the relations between the government and liberals like Vernadsky during July and August 1904. This phase began with the asassination of the hardline minister of the interior, von Plehve, on July 15 by a member of the Socialist Revolutionary party. Von Plehve was replaced by the reform-minded Sviatopolk-Mirsky, a relative of one of the leaders of the Union of Liberation, Dmitrii Shakhovskoi.[71] Shakhovskoi established close contact with his relative, who requested from zemstvo activists a memorandum with their ideas on reform. In the fall of 1904 a group of liberals, including Vernadsky, met with the minister of the interior and his chief assistant, S. E. Kryzhanovsky, who had been a student radical and a member of the Oldenburg circle at St. Petersburg University in the 1880s before joining the government.[72] Kryzhanovsky was to play an important role as a reform-minded bureaucrat over the next several years, drafting much of the legislation on representative institutions, civil liberties, and land reform between 1905 and 1907, as he served successive interior ministers such as Sviatopolk-Mirsky, Bulygin, Witte, and Stolypin.[73]

While conducting private talks with the new minister of the interior, the zemstvo constitutionalists and members of the Union of Liberation increased their public pressure on the government for reforms, realizing that the war with Japan was not to be ended soon and the government was therefore more vulnerable to pressure. In October 1904, a congress of the Union of Liberation was held which defined a concrete political strategy: to call a national congress of zemstvo representatives at which a resolution would be passed to create a national representative institution; to begin in November and December a national banquet campaign to agitate for reforms; and to form at these banquets various unions of professional groups like doctors, lawyers, teachers, engineers, university professors, etc. and unite them in a nation-wide Union of Unions which would work for reform.[74]

This strategy proved highly successful. The zemstvo congress held in November supported the program of the Union of Liberation; the banquet campaign of November-December attracted thousands of professional people; and between the fall of 1904 and spring of 1905, a series of professional unions were organized to agitate for reform. By May 1905 they had been loosely united in a national Union of Unions, headed by Paul Miliukov, a professor of history at Moscow University and one of the early members of the Union of Liberation. Vernadsky was to play a key role in the organization of one of these professional organizations, the Academic Union, which by the fall of 1905 claimed over 1,500 members, more than half of all the teachers in Russian higher education.[75] Their task was greatly facilitated by Sviatopolk-Mirsky's decree of December 12, 1904, which relaxed censorship, promised improvements in civil liberties and a limitation on the arbitrary actions of government officials.[76] Mirsky had also included a promise of a national representative institution, but this part of his draft was inked out by the Tsar.

Just a few days after Mirsky's decree relaxing censorship, Vernadsky and his associates, Sergei Trubetskoi, Ivan Grevs, and Konstantin Timiriazev published articles calling on Russian professors to end their traditional passivity and organize themselves to push for reforms, both in higher education and in society at large.[77] In his article of December 18, for example, Vernadsky called on his colleagues in higher education to hold a national congress of professors in order to dismantle the police regime in education, and to fight for academic freedom and autonomy and thereby save Russian universities from destruction at the hands of the Tsarist government.[78] These articles found a widely favorable response among liberal faculty members, but they also helped to polarize the faculty into liberal and conservative factions during the course of 1905–6, a polarization which deeply affected academic politics for many years afterward.[79]

These articles were published in an atmosphere of deepening crisis, both in Russian universities and in Russian society at large. A series of student demonstrations in St. Petersburg and Moscow during October, November, and early December, some with anti-war overtones, led to repression by the police, with a number of students severely beaten. On December 7, 1904, "a large meeting of Moscow University students voted to close down the university at least until after the Christmas vacation. Furthermore, the students asked the professors to protest against the brutality shown by the police."[80] The Faculty Council met on December 11 and appointed a commission of sixteen professors to look into the situation, in defiance of the minister of education's specific prohibition against such action. Three days later, this commission made its recommendations, including amnesty for the students involved in the December 7 meeting and dissolving the faculty disciplinary court, in light of Glazov's circular of the previous September which weakened its authority over students. The Faculty Council accepted these recommendations and "then declared that any reform of the universities was impossible without some measure of general political reform as well."[81] By the following February, faculty councils throughout the country followed suit with similar resolutions linking university reform with general political liberalization.[82]

By the time students returned from their Christmas vacations, the events of Bloody Sunday had occurred, and the confrontation between the Tsarist government and ever larger numbers of Russian subjects had deepened. One of Vernadsky's former students was killed in St. Petersburg on Bloody Sunday, apparently an innocent bystander of these events, shot twice in the back as he walked, briefcase in hand, through the Alexander Gardens. Vernadsky had considered this student, B. A. Luri, one of his most talented undergraduates and had invited him to remain in the university to prepare for a career as a professor when he received his degree in 1901. Luri had chosen instead to become a field geologist, preferring work outdoors to laboratory work and teaching. But he kept in close touch with his former teacher and had just met with Vernadsky during a trip to Moscow for the holidays. Vernadsky published an angry article in a leading liberal newspaper about this waste of human talent, declaring that "one more victim has fallen in the long martyrology of the Russian intelligentsia."[83] The moral indignation caused by such acts of violence was the catalyst for many previously quiescent professional people to join liberals and radicals in the coming months as they organized to force changes upon the government.

One of the ironies of Bloody Sunday is that it was used as a pretext to fire the liberal minister of the interior, Sviatopolk-Mirsky, and replace him with a more conservative bureaucrat, Bulygin. Bulygin tried to coopt change by ordering his ministry to draft a proposal for a purely consultative national assembly. The French Ambassador commented on the firing of Mirsky, who had been made the scapegoat for Bloody Sunday: "The era of liberalism that had barely opened is already closed."[84] In fact, however, his remark was a bit premature. The heyday of the liberals, though shortlived, was yet to come.

Students returned to university campuses in February just long enough to vote overwhelmingly for a strike until September. Despite the attempts of the Tsarist government to reopen higher education, Russian universities and higher technical schools were shut down for the next year and a half, with the exception of a one month attempt to reopen them in the fall of 1905. Despite their opposition to a student strike, which they considered counterproductive, most liberal and moderate professors opposed any attempts to reopen the universities by force, since they believed such action in an atmosphere of confrontation would lead to violence between students and government forces. Conservative professors, who were in the minority in most institutions at this time, dissented from this view and called for an immediate resumption of classes. Judging from discussions in faculty councils and in the newly organized Academic Union which held its first national congress in March, what the majority of professors wanted at this time, as a minimum, was autonomy for the universities and other institutions of higher education and general political reform which would include a guarantee of civil liberties and a national representative institution with real powers.[85] They disagreed with the view of the Tsarist government that they were mere civil servants (*chinovniki*) whose job was to prepare other middle-level officials for government service. Instead, they saw their role as the protectors, disseminators and creators of science and culture (*nauka*), in service to Russian society, not as servants of the central government.

In March 1905, Vernadsky was elected to the central council of the new Academic Union. He clearly expressed his viewpoint in a report presented at a conference of Moscow professors held in April. Vernadsky defended the Academic Union's call for granting the universities the right to elect their own officials and to assume responsibility for student discipline as a necessary step toward eventual full self rule. He reiterated that complete autonomy for higher education would only come with the creation of a national representative political system for the country as a whole, which could act as a guarantor of such rights. In the meantime, the professors had a special moral responsibility:

> We professors are in a special position. When we discuss events we cannot do so solely in our role of Russian citizens. We must also act as the guardians of the interests of science and knowledge. . . . Our first duty is not to let the higher educational institutions suffer during this period of social upheaval.[86]

The basic purpose of the Academic Union, Vernadsky maintained, was to unify the professoriate in order to protect higher education in an era of social upheaval. Neither the Tsarist government nor a largely uncultured general public could be counted upon to do so, Vernadsky believed.

Vernadsky's speech was made in the wake of the disastrous defeat of the Russian army by the Japanese at Mukden in February; this was reflected in this April comment:

> For centuries the government has regarded knowledge as a necessary evil. And now as a result of this policy Russian blood is flowing in the Far East. . . . The government began this war without any understanding of the capabilities of the enemy. . . . In the West and in Japan there is a general recognition of the power of science and knowledge and this underlies all state policy. . . .[87]

Over the next six months, there was a great deal of sparring back and forth between the professoriate and the central government. The government carried out individual reprisals against some teachers in higher education, firing a few, transferring others, forcing others to retire. The Academic Union sought to defend such teachers and publicize their cases.[88] In the Russian system, even senior professors were unprotected by any form of tenure, so that each act of reprisal tended to increase the resentment of liberal and moderate professors and consolidate their support for the Academic Union. In April, the official newspaper of the central government warned professors that they and all their students would be suspended if classes did not resume in September.[89] The government added to this a threat to hire replacements for Russian professors in Germany and other parts of Europe. Professor Koltsov, commenting on this threat in a letter to Vernadsky in May 1905, suggested that the Academic Union write to leading German academic publications about the Russian crisis and warn them of any possible attempts by the Ministry of Education to recruit professors abroad.[90] Koltsov's suggestion was carried out, and the warnings of the government along these lines proved to be empty threats.[91]

Vernadsky commented on the situation of the universities in several articles published in leading liberal journals during the summer of 1905.[92] He blamed the student strikes on the repression, arbitrariness, and violations of human dignity perpetrated by a badly organized government. If students resorted to strikes and even violence, it was because there were no peaceful, legal outlets to protest abuses or bring about change in Russian society. The threat by the minister of education to fire all professors while the student strike lasted, according to Vernadsky, "did not frighten us but deeply insulted us," as did the threat to hire replacements from Germany. These threats not only offended the national feelings of the Russian professoriate, they also reflected very badly on one hundred years of efforts by the Ministry of Education to build a system of higher education using talent and expertise from within the Russian Empire.

Vernadsky saw several ways out of the current impasse: calling a national congress of professors to write a new university charter to replace that of 1884, giving powers of self-rule to local faculty councils, and the creation of private universities, free from government control.[93] The latter proposal was apparently not original with Vernadsky but was discussed frequently in the spring and summer of 1905 at various meetings of liberal groups. For example, at a conference of zemstvo constitutionalists, held in Moscow in July 1905, liberal activists like Prince Petr Dolgorukov and Iu. N. Novosiltsev called on local zemstvos to assign some of their funds to help create such private universities and provide scholarships to students wishing to attend them.[94] At the second congress of the Academic Union, held in August, Professor E. N. Trubetskoi of Moscow University called on that organization to support the idea of creating "free universities." As Trubetskoi put it: "A free university would be a valuable supplement to the state institutions. . . . They would be more flexible since they would not be bound by the government teaching plans and the fixed distribution of faculty chairs. . . . Free universities would advance higher education by introducing new subjects and new methods of teaching."[95] Perhaps more importantly, should the state universities be closed by the central government in a period of national crisis, private universities could temporarily replace them.

The Academic Union approved a resolution supporting Trubetskoi and called on the Moscow and St. Petersburg chapter to prepare a detailed project for a university in time for the next congress of the Academic Union. The local chapters of the Academic Union were to secure support from the zemstvos and the municipal councils.[96] Such discussions eventually led to the creation of more than twenty "free universities," including Shaniavsky University in Moscow, which opened its doors in 1908, supported by money from a wealthy industrialist, by tuition fees, and by subventions from the municipal government of Moscow.[97] Vernadsky's student Fersman became a professor at Shaniavsky University soon after it opened, and in 1912 he offered the world's first university course on geochemistry in the "free university," where the curriculum was not subject to approval by the Ministry of Education.[98]

By the summer of 1905, Vernadsky had become such a well-known spokesman for the universities that a leading liberal publisher, I. V. Hessen, asked him to write a small book on the subject, to be part of a series on various problems ("the agrarian question," "the labor question," "the women's question," etc.).[99] Such a book was eventually published in 1909, written not by Vernadsky but by his friend and younger associate, the biologist N. K.

Koltsov.[100] In the meantime, the crisis of the universities came to a head in the summer and fall of 1905, and the liberals in the universities won what turned out to be something of a pyrrhic victory. The faculty were granted the right to elect their own rectors and deans, and student discipline was once again turned over to faculty courts.

This outcome was due to pressure from the faculty and to a serious split within the central government over how to deal with universities. The Ministry of Education itself was split over this question. For example, Minister Glazov took a hardline toward the entire faculty, whereas his subordinate, Nekrasov, the curator of the Moscow educational district, was willing to defy the minister and hold formal talks with the faculty, believing that most of the faculty wanted to resume classes and could be wooed away from the liberation movement, with its demands for general political reform, by timely concessions on university governance.[101] An even more serious split within the central government on this question surfaced in August of 1905 between the Ministry of the Interior and the Ministry of Education. The Tsar ordered a conference of his ministers to consider whether the universities should be reopened on September 1.

The conference opened amidst a bitter dispute between Glazov and Trepov, the Governor-General of St. Petersburg. Glazov had asked Trepov to ensure police cooperation in maintaining order in the universities if and when they reopened in the fall. But Trepov, in a surprise move, declared that the police would no longer enter the higher educational institutions. He informed the conference that in his view the basic problem was with the past policies of the Ministry of Education towards the universities. The police could no longer be expected to rescue the Ministry of Education from the results of its mistakes. Past experience had shown that repression only aggravated student unrest. Trepov noted that the Ministry of Education had had nine months during which the universities were closed to develop a new policy toward higher education but had failed to do so.

Trepov made two proposals to the conference. First he urged withdrawing the threat of April 16 to fire professors en masse if they failed to return to the universities in the fall. In view of the acute shortage of qualified academic personnel such a policy would only lead to the destruction of higher education in Russia. Secondly, Trepov suggested that the government issue temporary rules for the governance of the higher educational institutions until a new university charter could be presented to the Bulygin Duma, which the Tsar had just promised to convoke. Trepov argued that the government had nothing to lose. Matters could not get any worse in the universities. Perhaps if the professors were given more responsibilities they would be able to keep order there.[102]

Trepov's position was supported by most other ministerial officials at the conference. It was felt that the government needed to reopen the universities to protect its prestige. Professors should be given responsibility for maintaining order in these institutions. "In the event that the universities were forced to close again it was important that the onus be borne by the professors, not by the government."[103] On August 27, the Tsar approved new "temporary rules" for the universities and other higher education in Russia, rules which in fact were to last until the collapse of the regime in 1917. The faculty councils regained the rights they had possessed under the earlier charter of 1863 to elect rectors and assistant rectors, and individual faculties regained the right to elect their own deans. Faculty courts became responsible for student discipline and the faculty councils were made responsible for overall order on the campuses. These rules were supposed to last until the Duma replaced the charter of 1884 with a new piece of legislation, an event which never occurred.

Essentially, the temporary rules of August 27 were never meant by the government to be a fundamental reform of higher education. They were meant to be a political concession to the Russian public which would help restore an atmosphere of normalcy to the country. The temporary rules followed the signing of a Peace Treaty with Japan and the Rescript of August 6 [the Bulygin Duma]. They were regarded, as the Conference of August 11 shows, as an essentially tactical move in the government's overall strategy of regaining control over events. By the end of August it seemed that the wave of political turbulence was on the wane. The labor movement was relatively quiet. The Bulygin Rescript had split the liberal camp. The revolutionary parties, at least in Central Russia, had not yet shown that they enjoyed significant mass support. Yet, in retrospect, it is absolutely indisputable that the issuance of the temporary rules of August 27 was a major miscalculation. Even Witte and Lenin agree on that point. The rules ushered in a new phase in the Revolutions of 1905.

While the government was discussing what to do about the universities, the professors themselves were meeting at the second congress of the Academic Union, which convened in Moscow on August 25. Vernadsky gave the main report from the Moscow branch of the union, while his friend from student days, Professor I. M. Grevs, gave the report of the St. Petersburg branch.[104] Despite the police harassment of a number of prominent professors earlier that summer, by late August the majority of the moderate and liberal professors were seeking an accommodation with the government, an arrangement which would allow the universities to reopen in September.

Vernadsky became the chief spokesperson for returning to the classroom.[105] He argued that by resuming their teaching roles, they could be able to use the universities as a public forum and increase pressure on the government for general reforms. He also felt that "if the government granted freedom of speech and of the press, it was likely that the student movement would lose its former importance and political tensions would move outside the universities."[106] As it turned out, Vernadsky could not have been more wrong. However, the second congress of the Academic Union supported his call for a return to teaching by an overwhelming majority of the 129 representatives in attendance from thirty-nine different universities and higher technical schools. The final resolution insisted only that the faculty, protected by the power of the Academic Union, would not give in to government threats and would cease teaching if the government once again resorted to police repression against students.

These moves proved to be a major miscalculation on the part of both professors and the government. On the one hand, the government agreed to give the universities the autonomy professors had for so long demanded, on condition that they take responsibility for maintaining order on the campuses. The government would no longer use the police and troops to keep order there. This suited the Academic Union's own position very well. They were against a police presence on the campuses and believed that once the faculty had been given autonomy, it could maintain order on its own. However, having foresworn police protection against student disturbances, the universities proved defenseless once students returned in September in a mood more angry and militant than anyone had predicted.

The more radical student organizations seized the opportunity presented them and invited workers and revolutionary parties to make the universities "islands of revolution." During the day, campuses would be crowded with students attending classes, but at night as many as thirty thousand workers and others would flock to the universities, which were transformed into a kind of town meeting, the only place in Russian society where freedom of assembly existed at the time. Much of the senior faculty, Vernadsky among them, feared that this would bring down upon the universities the wrath of the authorities and threaten their existence. But when university officials and the faculty pleaded with students to return to their studies and not allow the universities to be a forum for political controversy, the students ignored them. Whereas the liberal faculty had won a number of major concessions from the government, such as university self rule and the promise of a national Duma, students still felt in a confrontational mood, demanding a variety of other concessions, including a share in running the universities, increases in financial aid and changes in the way aid was administered, the return to an elective course system, removal of restrictions on the admission of women and Jews, and so on. Their sympathies were with disaffected groups outside the universities, and they rejected the pleas of the faculty not to invite outsiders onto the campuses.

At Moscow University, the "Faculty Council was unwilling to accept the continuation of the popular meetings, which became even more numerous with the outbreak of the Moscow Printers Strike on September 19."[107] As it turned out, the printers' strike marked the beginning of a nation-wide general strike in October. On September 22, the Faculty Council voted to close the universities. This left many of the faculty uneasy, however, since it sent the wrong message to the government, which was already highly skeptical about the ability of the faculty to keep order in the universities. At another meeting of the Moscow Faculty Council, held on September 24, Prince Trubetskoi, the new rector, recommended that the university remain closed "until the workers quiet down."[108] Vernadsky advocated a somewhat different strategy, which turned out to be equally ineffective. "Vernadsky recommended that the Faculty Council directly and publicly petition the government to allow freedom of assembly in Russia. If that were granted, the universities would hopefully be spared further direct involvement in the political struggle." Vernadsky hoped that public meetings would be held elsewhere and the universities spared the ensuing turmoil.[109]

The Faculty Council adopted Vernadsky's suggestion, and on September 29, the rector, Prince Trubetskoi, traveled to St. Petersburg to petition the government for nationwide freedom of assembly. While discussing this and other issues with the minister of education, Trubetskoi suffered a massive heart attack and died on a couch in the minister's office.[110] Trubetskoi's newly elected assistant, Professor A. A. Manuilov, succeeded him, and Vernadsky was elected assistant rector, a post he filled until April 1906, when he was elected as a representative of the universities to the State Council, the upper house of the newly constituted legislative branch of government.[111]

As a general strike gathered momentum throughout the country in early October, street violence between left wing and right wing crowds increased, often with the universities as the locus of action. With hundreds of thousands of industrial workers and railway men on strike, the resulting economic disruption brought a rightwing reaction from small-scale merchants, artisans, butchers, and other tradesmen, who had traditionally been monarchist and under the influence of the police, which regulated their activity and granted their licenses to operate. Fearing an increase in disorders, a special conference of government Ministers on October 13 ordered the faculty to keep non-students away from the campuses.[112] However, that did not prevent a mob of butchers and milkmen from invading the campus of Moscow University on October 15, intent on beating up students and members of minority groups like the Jews, whom they blamed for the economic losses they suffered as a result of the general strike.

In an effort to quell the violence and end the strike, on October 17, the Tsar signed a manifesto granting civil liberties and a wider franchise in the upcoming elections for the new Duma. Most importantly, unlike the Bulygin Duma promised in August, which was intended to be purely consultative, the October Manifesto promised that no new laws were to be enacted without the approval of the Duma.[113] These concessions split the opposition, with some of the more moderate elements, particularly among the professional and propertied classes, now willing to work with the government against the revolutionary opposition. Frightened by the increase in violence, both in rural and urban areas, these elements strengthened the hand of the government as it turned from concessions to repression during the fall and winter of 1905–6. Street demonstrations by students and others waving red flags, singing revolutionary songs, and shouting militant slogans were followed after the October Manifesto by attacks carried out by right-wing gangs—the Black Hundreds—against students, intellectuals, and national minorities.

On October 20 in Moscow, a massive funeral procession to honor a fallen radical, Nikolai Bauman, was broken up by Cossacks. Following the attack, students seized Moscow University and erected barricades, but on October 28, they gave in to the demands of the military governor of Moscow and cleared the university in return for a promise that the Cossacks would be withdrawn from Moscow.[114] In this atmosphere of street violence, the Faculty Council of Moscow University met on November 7 and voted to keep the university closed until the start of the second semester, explaining that "the faculty council has an obligation to the nation to preserve the University as an institution of enlightenment."[115]

Since classes were suspended during nearly his entire six-month tenure as a university administrator, Vernadsky spent much of his time the first few months trying to protect the university's physical resources from attacks by the right and the left. Gangs of right-wing toughs, stirred up by rumors that the Tsar had fallen under the control of Jews and foreigners, invaded the campus of Moscow University on October 15 and again in the wake of the Tsar's proclamation of October 17.[116] Fedor Oldenburg was nearly killed in an attack on liberals by Black Hundred gangs in Tver, who burned down the zemstvo building after they heard about the October Manifesto.[117] Invasions by the Black Hundreds were soon followed by a takeover of the campus by the left. Vernadsky spent much of his time pleading with students, workers, and revolutionaries who had commandeered university buildings not to wreck the valuable physical resources of the university.

The situation became critical in December when the Socialist Revolutionaries and Social Democrats prepared an uprising in Moscow, using the university as headquarters for more than 2000 armed men, mostly radical students and workers. As his son George Vernadsky later expressed it, "I remember vividly what despair my father and the professors of chemistry and physics felt when a group of student fighters broke into the laboratories where experiments of great scientific significance were being conducted. Revolutionary students were preparing to make this their staff headquarters and a transit point for the wounded. Only with great difficulty did the professors succeed in talking them into occupying other university buildings for their purposes."[118]

By December 18, the military had suppressed this rebellion at the orders of the new prime minister, Sergei Witte. But Witte decided not to reopen the universities until the fall of 1906. In a personal meeting with Witte held in February 1906, the rector of Moscow University, Manuilov, and his assistant, Vernadsky, were told that the government did not want a large concentration of students in the cities during the upcoming elections and the opening of the Duma.[119] With the university closed, Vernadsky spent much of this year in a variety of political and philanthropic activities.

Starting in October, he helped to organize the Constitutional Democratic party, which became the largest opposition party in the country to participate in elections to the first Duma. Vernadsky's apartment became a center of the party's Moscow activity.[120] In fact, as early as the previous July, in a meeting of zemstvo constitutionalists, Vernadsky had joined in calls to form a liberal political party, considering the Union of Liberation too broad and too amorphous in its program. Now that the government had promised some form of national Duma, it was urgent. Vernadsky argued in July, to form a political party with a detailed platform. Otherwise, he feared, the "popular movement will bypass us. We have few adherents because of the lack of a program and organization. We do not yet have a realistic economic program, workers' program, financial or other programs."[121]

Vernadsky's fears about being bypassed by the popular movement, however, proved premature. The Kadet party which he helped form and on whose central committee he served from 1908 until 1918 soon emerged as the largest of the non-revolutionary opposition parties in Russia. Of the more than one hundred different political organizations which came into existence during the first election campaign, only four had a nationwide organization. The Kadets, with between 100,000 and 120,000 members by the spring of 1906, dwarfed their nearest rivals, the Octobrists. The latter had somewhere between 10,000 and 24,000 members in the same period, while the Trade and Industry party and the Party of Legal Order were even smaller. Only the Kadets and the Octobrists had at least one party committee in most Russian provinces.[122]

Bankrolled by a small group of wealthy noblemen, the Kadet's leadership was made up primarily of professional people (half of the central committee were university professors or former professors) and liberal gentry. Only one or two industrialists could be found among the leadership. Commercial and industrial groups, which had a tradition of political apathy in Russia, were attracted to rivals of the Kadets such as the Octobrists or the Trade and Industry party, if they joined a party at all. The Kadets not only had the largest number of members among non-revolutionary opposition parties which participated in the elections to the first Duma (the revolutionary parties for the most part boycotted the elections), they commanded an impressive array of newspapers, journals, and book publishers. In addition to newspapers directed at the educated classes, they published one of the first national papers directed primarily to the peasantry, *Narodnoe delo*, published in Moscow. This was one way in which the party hoped to attract mass support from the peasantry. In all, the Kadets published between forty and fifty local newspapers in this period, more than any other political party.[123]

Vernadsky sat on the editorial boards of several of these publications; and while assistant rector of Moscow University, he also tried to regain control of Moscow University's press, which had been leased in the 1870s by the university administration to the ultra-conservative Moscow publisher, Katkov.[124] In these months of political turmoil in late 1905 and early 1906, the Kadets clearly tried to extend their cultural hegemony as much as possible; and Vernadsky, a close friend of the party's first two secretaries, Dmitrii Shakhovskoi and A. A. Kornilov, played a key role in these efforts. This is a role which later Soviet historians, trying to appropriate Vernadsky's scientific reputation for their own purposes, have tended to down play, particularly given the antipathy of Lenin and the Bolsheviks toward the Kadets.[125] Soviet historians have in recent years made a distinction between sincere but misguided liberals like Vernadsky whom they see on the periphery of the Kadet party, and traditional liberal "villains" such as Struve and Miliukov whom they have portrayed at the center of the party as representatives of the bourgeoisie. Such views need serious revision, since scientific intellectuals like Vernadsky were clearly much more important to the party's early years than Soviet scholars have indicated; and the party itself in these early years was much less a center of the commercial and industrial bourgeoisie than it was of an educated professional elite and liberal agriculturalists.

Among his other political and philanthropic activities during the winter of 1905–6, Vernadsky and his son George also helped to organize famine relief for the peasants of Morshansk county who were once again suffering from hardship brought about by the bad harvest of 1905.[126] In the process, George Vernadsky, a student of history at Moscow University, was arrested together with the overseer of his father's farm, A. I. Popov, and Popov's son. In providing famine relief, George Vernadsky had been accompanied by the Popovs, who were peasants themselves who knew the local people, and by the local priest and four school teachers. What aroused the suspicions of the authorities was that this group not only provided famine relief but explained to the peasants the upcoming elections to the Duma and urged them to vote. As George Vernadsky recalled in memoirs written many years later, their propaganda among the peasants was not revolutionary but was sympathetic to the radical Peasants' Union. The union called for a redistribution of lands from the gentry to the peasantry, a position supported by many Kadets such George's father, provided it was done in an orderly fashion and with compensation paid to the former owners.

After several weeks of this activity, George Vernadsky and the Popovs were searched and arrested by the police, who were seeking revolutionary propaganda. Although they found none, George spent a week in jail. When his father learned of his confinement, he telephoned his former schoolmate at St. Petersburg University, S. E. Kryzhanovsky, who was now assistant minister of the interior. Thanks to Kryzhanovsky's intervention, George was released, but the Popovs remained in jail.[127] Therefore, on a visit to St. Petersburg in early February on university business with Manuilov, Vernadsky brought up the matter during an interview with Count Witte. Witte promised to intervene and secure the release of the Popovs, provided Vernadsky agreed not to use the affair for anti-government propaganda.[128] Vernadsky provided the minister with a memorandum to that effect, and the Popovs were soon released.[129]

In retrospect, the arrest of George Vernadsky and the Popovs was not an isolated incident but was part of a campaign of government harassment against newly formed opposition parties, such as the Kadets. As Emmons has noted in a recent monograph, "governmental interference in the political activities of the Kadets and other parties in the winter and spring of 1906 was widespread."[130]

According to Emmons, arrests of Kadets during the election campaign occurred in at least twenty-four provinces, and the authorities were especially concerned about contacts with the Peasants' Union. The police often targeted local kadet party committees, arresting their members, searching their apartments and prohibiting the publication of party literature. Many were held in jail during much of the election period on a purely administrative basis, without formal charges being brought or trials held.[131] A circular of January 20, 1906 from the minister of internal affairs emphasized that local governors and police chiefs should be very cautious about allowing any kind of public meetings.[132] Political parties had not been officially recognized yet by the government and therefore had a shaky legal status throughout the first election campaign, a fact which did not prevent the Kadets from winning control of forty percent of the seats in the first Duma, when it convened on April 27, 1906.[133]

In general, the gap between the government and a growing number of the propertied elements of Russian society on the one hand and the Kadets and their allies on the other grew considerably between the fall of 1905 and the spring of 1906. The government's turn toward repression and its harassment of opposition political parties increased the skepticism among opposition elements about the sincerity of the government's reform efforts and the durability of its concessions toward a constitutional system based on law and respect for individual rights. The government and many propertied elements, especially among the gentry, were for their part increasingly suspicious of groups like the Kadets. Although opposed to revolutionary violence, Kadets supported the general strike in the fall of 1905 and made it official party policy to court votes among the peasants, supporting in its platform forced expropriation of land.

The Kadets also sought support among the national minorities, especially Jews, Muslims, and Poles, some of whose organizations had close ties with the Kadet Central Committee. In fact, nearly a quarter of the members of the Kadet Central Committee were Jewish, a fact likely to arouse antipathy among ruling circles, where antisemitism ran deep. Although the other three quarters of the Kadet Central Committee were, in fact, Russians of mostly noble origins and the Kadet party was dedicated to increasing Russian national strength on the basis of peaceful change, the rising tide of reaction among the landed gentry and propertied urban elements made the Kadets, as the largest opposition party, easy scapegoats for the violence.

As a result, the Kadets after the fall of 1905 rapidly began to lose influence in the zemstvos and universities which had until recently been their institutional strongholds. In the countryside, the conservative tide in the zemstvos was occasioned by an increase in peasant attacks on gentry estates, becoming particularly severe after the October Manifesto. Most of the attacks were on the property, rather than upon the persons of the landed nobility; they were usually well organized by local villagers, but they tended to be very selective. The peasants involved usually attacked the largest estates, those over 1500 acres, which were often the estates of their former owners. Only about ten percent of peasant villages were involved in such attacks; the majority of peasants at the time apparently preferred peaceful, legal change; and they turned out in record numbers to vote for the first two Dumas, which they hoped would carry out such land reform. Just as the Duma began to debate the issue of agrarian reform, Vernadsky, who spent part of June 1906 at his estate in Morshansk county, wrote his wife that the mood in the local countryside was quiet but nervous. He thought that newspaper reports of peasant disturbances were exaggerated, and that most of the disturbances were focused on a few targets. His own estate apparently escaped trouble.[134]

Whatever the facts of the case, panic seized much of the landed gentry after October 1905. Some fled the countryside, others tried to sell their lands or posted armed guards on their estates. By the spring of 1906 their panic had turned to political reaction, most readily visible in the organization of the United Nobility, a group pledged to work with the autocracy to preserve the privileges of the nobility and protect their property rights in the countryside. Elections to the zemstvos in this period led to the ouster of the liberal gentry who had played such an important role in these institutions over the previous several decades. The Kadets' support of the general strike in the fall of 1905 and their courting of the peasantry with promises of agrarian reform may have made them the single largest party in the first Duma, but it lost them the support of much of the local gentry who held the political balance of power in the countryside. With their allies among the liberal gentry largely driven from power in the countryside, the zemstvo professionals—physicians, teachers, statisticians, and agronomists—were fired by the thousands. Zemstvo budgets, especially for public health, education, and other social services, were slashed by as much as one-third to one-half in 1906 and 1907.[135]

Peasant violence resulted from a number of factors which had steadily accumulated over the years, such as pent-up land hunger and population growth, high taxes and redemption payments, and increased cultivation of rural lands by the gentry themselves. More immediately, however, the bad harvest of 1905 brought many peasants to the brink of starvation, particularly in the provinces of Central Russia.

Despite attempts by social activists like Vernadsky and his friends to raise private relief in early 1906, such efforts proved much less effective than in 1891. The government, in the grip of its own financial crisis, was also unable to provide as much relief as it had provided in 1891. Not until June 1906 did the new Duma send a bill to the State Council assigning 15 million rubles for seed and produce to provinces hit by bad harvests and famine.

The Duma bill caused lively debate in the State Council, to which Vernadsky had just been elected as a representative of the universities. The minister of finance, Kokovtsev, spoke against the bill, citing its costs and the fact that the state budget was already in arrears. This led to a heated debate between Kokovtsev and Vernadsky. The minister accused Vernadsky of distorting his words and using famine as a political issue in order to focus on the mistakes of the government rather than to feed the hungry. According to Kokovtsev, the state simply did not have the money and it was his duty to point out that fact. Vernadsky countered by arguing that the newly elected Duma from now on would ask for accountability concerning how the people's money was spent. According to Vernadsky, the minister of Finance should be seeking new sources of revenue to provide such necessities as famine relief rather than simply seeking to limit expenditures.[136]

Vernadsky's interest in agrarian problems went well beyond the question of short-term relief for famine victims. Vernadsky envisioned an entire program of measures, relying on a combination of land reform and the application of science to agriculture. Vernadsky drew his inspiration from the American models of family farms and state aid and the Prussian model of commercial estates, both of which were much in the minds of the landed gentry during these years. Vernadsky was subsequently appointed to the Agrarian Commission of the Kadet party in early 1906, and later confided, "I was a member of the left group in the Agrarian Commission."[137] Vernadsky's leftism does not seem particularly radical to later generations, although it proved unrealizable in the last years of the Tsarist regime. As Vernadsky confided both to his diary and in a newspaper article in May and June 1906, basic agrarian reform must accompany the political changes being introduced or the latter would fall. He criticized the Kadets for not paying enough attention in their party platform and elsewhere to the application of the latest achievements of science and technology to agriculture.[138]

"A resolution of the agrarian question has now become a necessity for the state," Vernadsky wrote in his diary in June 1906. Vernadsky's solution to the agrarian problem agreed with the populist (Socialist Revolutionary, or SR) platform calling for the redistribution of land to working peasants, but he disagreed with the SR emphasis on collective ownership of the land by peasant communes. Vernadsky felt that the land must, with proper compensation, be redistributed to the peasant tillers, but "obviously it is not permissible to stop with this one measure. Other practical measures must be carried out to improve agriculture." He then outlined a program of measures which the central government needed to adopt, such as the provision of regular credits to peasant agriculture, the development of a system of credit cooperatives among the peasantry, the organization of more schools to teach scientific agronomy, the organization of experimental stations and model farms, the development of a central weather service and the formation of a soil committee to advise farmers on how to maintain and increase the fertility of their land. These were issues he had thought about for years, both in his work with Dokuchaev, the father of Russian soil science, and in his experience as an estate owner. Like his friend Shakhovskoi, Vernadsky considered the populists' emphasis on reforming agriculture by relying on the peasant commune to be an illusion. "This in reality is a fantasy. The answer is in free cooperative societies."[139] Unfortunately, liberals like Vernadsky and Shakhovskoi were losing influence locally at precisely this time, and they had never enjoyed power or influence in the central government. The land reform program of Stolypin and Gurko, launched after 1907, was a far cry from Vernadsky's program of combining private initiative with science and government assistance. Although Vernadsky was appointed to the Agrarian Commission of the State Council in 1908, he was so isolated in that body of government bureaucrats and conservative landed gentry that his voice was simply ignored.

In the meantime, not all the peasantry waited patiently for land reform. Many of them survived the lean winter and spring of 1905–6 by plundering the estates of large landowners. One historian has estimated that the plunder amounted to as much as one hundred million rubles, a sum sufficient to make the difference between survival and starvation for many peasants.[140] Peasant unrest profited in other, more substantial and long-term ways as well. Between November 1905 and November 1906, frightened gentry sold off as much land to the peasantry as they had sold during the thirty-five years since the emancipation. Land prices in many areas dropped and rents also declined by some one-third. Redemption payments, which peasants had been required to pay both for the land they received under emancipation and for value as serfs, were abolished entirely in this period. The government abolished some taxes which fell exclusively on the peasantry and more nearly equalized others where the peasants had been paying an unequal share compared with other social groups.[141]

Along with concessions to peasants, the central government also responded to peasant violence with armed retaliation. Military courts-martial sentenced thousands to die. Between October 1905 and the convocation of the first Duma in April 1906, police and government troops shot some thirty-four thousand rioters, of which an estimated fourteen thousand died.[142] The shootings and executions of 1906 set off a bitter debate in the Duma and State Council over capital punishment. To halt these state-sanctioned killings, the Kadets and others in the Duma passed a bill outlawing capital punishment; however, government officials and frightened members of the gentry in the State Council blocked the bill. All proposed laws had to pass both the popularly elected Duma and the State Council. Half of the members in the State Council were appointed by the Tsar and the other half were elected from highly restricted curia representing mostly the conservative classes in Russian society. A small group of liberal representatives elected by the universities and the Academy of Sciences provided an exception to the regular composition of the State Council. Vernadsky was nominated by the Central Committee of the Kadet party to stand for election to one of six positions reserved for the universities. Vernadsky agreed reluctantly, writing his son George in March 1906 that he did not want to waste his time on what might not be a useful post.[143] Vernadsky submitted to party discipline and was elected. He quickly clashed with the majority of members and with Prime Minister Witte in a debate over the first Duma's call for a general amnesty. Shouted down several times by the conservative majority, he was finally allowed to speak on May 4, 1906, criticizing the government for its policy of repression and mass executions.[144]

Although disappointed with the effectiveness of the State Council, Vernadsky remained a member until expelled in 1911, when his resignation from Moscow University disqualified him for his elected position.[145] He participated in only one other debate, the June debate over the Duma's bill for famine relief. Vernadsky attended most sessions of the State Council for the next five years, but his role was primarily that of observer. He saw little chance of accomplishing much in this body, which he correctly perceived as a check on the more democratically elected Duma. Most members, he wrote to one of his former students, the geochemist Ia. V. Samoilov, were "convinced defenders of the military-police autocratic government. . . . In the State Council I saw these people, impoverished in spirit—yet power is in their hands."[146]

Vernadsky and the other oppositionists—twelve to sixteen representatives of the universities, Academy of Sciences, and a small group of Polish members—at first tried to play an active role in the State Council, unmasking its reactionary policies in the face of intense hostility from the conservative gentry and the bemedalled and gold-braided ranks of Tsarist officials who made up the majority of the council. Vernadsky wrote his son in late May 1906: "It is very difficult and very unpleasant in the State Council. I remain there out of party discipline and a realization in my mind (but not in my feelings) that it is possible in certain circumstances to be useful there. Doubtless, our small group (12–16 people) can now and then make a breach and hinder them from the wholesale introduction [of whatever they wish]."[147] A few days later, however, in an article for a liberal newspaper on the State Council, Vernadsky struck a much more pessimistic note. He declared that the representatives of the opposition in the State Council could realistically "do nothing, since they always find themselves in a milieu that is for the most part directly hostile toward them."[148] In the same article, Vernadsky argued that the State Council was organized to obstruct the Duma and keep the old regime in power. The State Council contained no representatives of the peasantry, of urban workers, the village clergy, or merchants, and only token representatives of the industrialists and the universities. The State Council, he declared, could not represent the interests of most Russians because it was dominated by the rich minority of the population. He believed it was deliberately created to prevent significant agrarian reform and improvement of the situation of urban workers. He foresaw, quite accurately, the creation of a deadlocked political system with inevitable and constant conflicts between the Duma and the State Council.[149]

In late June 1906, the Duma passed and sent to the State Council a law abolishing capital punishment. Vernadsky wrote to his wife: "Yesterday we debated the question of the death penalty and, my god, what a stir it created!"[150] According to Vernadsky, a veritable Black-Hundred mood, that atmosphere of a lynch mob, reigned in the council. On June 28, Vernadsky was appointed to a special commission of the State Council to consider the Duma law abolishing capital punishment, but he was the only supporter of the law among the fourteen members. The commission and the State Council subsequently rejected the Duma's proposal. In his letter to his wife of July 2, he saw an urgent need to reform the State Council, since it was having a more and more detrimental effect on reform attempts.[151] Vernadsky expressed his indignation in an article of July 10 in *Rech,* one of the largest liberal newspapers. He was very critical of the government for its failure to observe due process and its reliance on a "white terror" of shootings and hangings to restore order. Such policies, he declared, were not worthy of a civilized country, and he asserted that the abolition of the death penalty was a "great service to science and scientific thinking."[152]

Vernadsky's willingness to invoke science was more a polemical device than a piece of argumentation backed up by evidence, since he did not cite any scientific research on the death penalty. Rather he invoked science as allied with civilizing forces which would make the death penalty unnecessary. The implication was that all right-thinking people who wished to be on the side of science and civilization would line up against the repressive policies of the Tsarist government. As a well-known scientist himself, he was clearly drawing the battle lines, invoking the prestige of "science" on his side and attempting to mobilize the forces of moral indignation against the government. Among the educated professional classes, such rhetoric no doubt had some effect, but among the landed gentry and bureaucratic officialdom the fear of revolution and determination to restore order at whatever cost clearly put liberals like Vernadsky beyond the pale.

Even among the educated, particularly among those with positions to protect, liberals found their support waning by 1906. The liberals did not help their own situation by grandstand gestures. For example, when the Tsar dissolved the first Duma on July 8, 1906, an angry mob of Duma members adjourned to the relative safety of Vyborg in Finland to discuss what they should do next. The far left wanted the group to make a mass appeal to the population for armed rebellion. Miliukov and the Kadets proposed instead a compromise resolution which appealed to the people to refuse to pay taxes or provide recruits for the army.[153] Vernadsky, who journeyed to Vyborg as a member of the Kadet Central Committee and carried the appeal back to St. Petersburg but apparently did not sign it, wrote to his wife soon afterwards: "I hold the opinion that this was the first step . . . I consider this a completely correct phenomenon of passive resistance, completely necessary in these moments of history."[154]

While the Vyborg appeal may have made its signers and other oppositionists feel better at the time, it was in fact a gesture that cost them considerably in influence. The signers of the appeal were not only arrested and sentenced to three months in prison they were also barred from future political activity, including running for office in the Duma. By this stroke much of the cream of the Kadet party, for example, was barred from serving in subsequent Dumas. This fact, combined with efforts on the part of the Kadets to tone down peasant demands for land reform, cost them dearly in support. The final blow came with the Tsar's new electoral decree of June 3, 1907, which created a system of weighted voting through curia which gave the predominant political influence in the Duma to some 20,000 voters from the landed gentry. The increasingly conservative gentry, therefore, came to dominate the zemstvos and the State Council, and after June 1907, the Duma as well.

The Kadets were never again as influential as they were briefly in 1905 and especially during the first Duma, from April to July 1906. The Kadets lost their institutional bases in the zemstvos and to some extent in the universities, and the machinations of the government and their own mistakes undermined their position in the Duma and in the peasantry. Among the urban classes they had to compete with more radical parties which often appealed more successfully to urban workers and educated but propertyless professional people. The Kadet's bid to become a powerful, mass party was largely unsuccessful after 1906. They remained influential in the print media, in education, and in some city governments like the Moscow Duma, where they were allied with progressive merchants and industrialists who generally belonged to other parties. But they failed to build the mass base which was the declared goal of leading members of the Kadet party like Vernadsky when the party was formed in 1905. While they succeeded in leading a movement which had forced some democratic concessions from the Tsarist government in 1905, their movement had left the central government embittered and more intransigent than ever, determined to roll back as many of the political concessions as possible. A barricade mentality of hatred and distrust characterized both sides—Tsarist officialdom and most of the gentry on the one hand, liberals like Vernadsky in the Kadet party on the other. Vernadsky's attitude of extreme distrust and hostility to the Tsarist regime that he expressed in this period both in his polemical articles and privately in letters was a reflection of the general distrust felt by Kadets toward the Tsarist regime. They clearly doubted the sincerity of the government in carrying out reforms.

Even in the universities, which together with the zemstvos originally had been one of the institutional pillars of the liberals, the swing to a conservative mood among the senior faculty was already noticeable by January 1906. Several factors accounted for this change in atmosphere. The strikes, demonstrations, and armed rebellions of the previous three months had seriously disrupted the universities and made classes impossible. Both the junior faculty and the students were asking for shared governance and an important role in running higher education, demands which a large sector of the senior faculty found threatening. Finally, a new minister of education, Ilia Tolstoi, was appointed at the end of October 1905.[155] Tolstoi actively sought an alliance with moderate and conservative faculty, and in January 1906, he convened a conference of elected representatives of the senior faculty from all Russian universities to help him draw up a new charter for university governance. The debates of this conference reveal that the liberal tide among the senior professoriate had already begun to wane. A slight majority of the conference voted that the

Ministry of Education should still have the power to confirm all professorial appointments, and an article sponsored by a liberal professor to the effect that "a professor is not considered a civil servant *(chinovnik)* and enjoys neither rank nor decorations" could not be passed. Some long-standing demands of the liberal professoriate were approved, however. For example, the majority of delegates voted that religion, ethnic origin, or sex should not be obstacles to becoming a professor. They also voted for the admission of women to the universities and the reduction of tuition fees. However, when it came to extending to the junior faculty voting rights in the faculty councils, the delegates overwhelmingly rejected such a proposal.[156]

Although the Tolstoi conference did not lead to a new university charter and Tolstoi himself was soon replaced with a more conservative minister of education, the deliberations of the conference were a clear sign that the tide in the universities had begun to turn against liberals among the professoriate. Another clear sign was the withering away of the Academic Union, which had originally been formed and led by liberals like Vernadsky. By early 1906, much of the senior professoriate had left the Academic Union, which had been taken over by the junior faculty. By 1907, the Academic Union was a mere shadow of its former self and had ceased to be an important factor in university politics.[157] There is no evidence that liberal senior professors like Vernadsky continued to participate actively in the union after the formation of the Kadet party in October 1905. The Kadet party absorbed an increasing proportion of their energies and refocused much of their political activity on national issues outside the universities.

After 1906–7, the government, particularly under Stolypin, moved to isolate the liberal senior faculty in the universities. The evidence suggests they largely succeeded. The government also moved increasingly to limit the autonomy of the universities, and much of the faculty, weary of conflict and eager to return to their professional roles as scholars and teachers, assumed once again the passivity that had characterized them during the 1880s and 1890s. Vernadsky remained politically active longer than most, but given three years of intense political activity (1904–7), in which he accomplished little in the way of scientific research, he too worried about his career as a scientist and was eager to return to it. As he wrote in his diary during this period, "a scholar regrets the time lost, the energies wasted without results . . . doubts never leave him: is this flickering of a candle worth the cost?"[158] His numerous articles on political themes and educational policy ceased by mid-1908 and did not resume until the university crisis of early 1911. Instead he concentrated on scientific expeditions, on developing the new field of geochemistry, mastering new equipment and methods for the spectroscopy of Russian minerals, and studying radioactivity.

The events of the 1905 revolution had left deep scars on the university, including a breakdown of collegial relations among faculty members with different political orientations. Matters reached a point where liberals and conservatives on the faculty sometimes refused to shake hands with each other during chance encounters in the dining hall.[159] Far more serious, however, was the inability of different factions to work together cooperatively and to present a united front to the government in an effort to protect the gains made during the recent period of unrest. Although university enrollments nearly doubled between 1904 and 1908, funding did not keep pace and the amount spent per student, including financial aid, which declined considerably.[160] The Tsar was determined that there should be no new universities, and the existing ones were overcrowded and underfunded. There was even some talk in court circles and the government about abolishing the universities altogether and concentrating on specialized technical institutes, on the theory that the latter would be more politically reliable. Although the universities were not abolished, the government did adopt a policy favoring technical institutes in construction of new facilities and general resource allocations. Such a policy aroused the ire of professors like Vernadsky, who believed in the broad scientific and humanistic education offered by the universities, but their articles on this subject had little effect on government policy.[161]

In 1908, the government under Stolypin began to attack the concessions made to the universities in 1905. "The Ministry of Education, acting under Stolypin's direction, began a policy of systematic counterreform."[162] In an article published in a leading liberal newspaper early in 1908, Vernadsky indicated that the universities had always been threatened by the extremism of the left and the right, but that the chief danger at the moment came from the right.[163] His assessment was well-founded. On New Year's day 1908, Stolypin fired his more moderate minister of education and replaced him with a man whom he advised to "be hard!" and to fire professors who did not cooperate in restoring order on the campuses. Underlying this policy of repression was an overestimation on Stolypin's part of the dangers of student activism and the ability of students to link up with workers. The events of 1905 loomed large in the memories of both Stolypin and the Tsar, but in fact, students were more divided than ever on political questions and less successful than ever in their ability to disrupt the universities or link up with other discontented groups in Russian society.[164] However, as Samuel Kassow has pointed out, the universities made a highly visible target for a prime minister like Stolypin, who was himself under pressure from the extreme right and could not afford to be "soft on students and intellectuals."[165] The universities were an even more inviting target because of their vulnerability and the divisions existing within them among the faculty and between students and faculty.[166] They had few defenses against a determined effort by the government to turn back the clock to the more repressive era when the charter of 1884 was in full force.

The new minister of education, Shvarts, began a campaign of counterreforms in the universities, which included increased Russification, reinstitution of quotas for the admission of Jews, exclusion of women even as auditors, restrictions on student organizations, and an attack on the new elective curriculum that Shvarts believed "allowed more time for revolutionary activities." He also wished to reinstitute a university inspectorate that would be independent of the elected rectors and faculty councils. All but the latter measure enjoyed some degree of success during his tenure.[167] Both Stolypin and Shvarts had an oversimplified view of the students and faculty, seeing both groups as allied in a common revolutionary effort, ignoring the great number of divisions actually existing among both groups. But Shvarts's attempt to force a new restrictive university charter through the Duma failed, dooming at least some of his most severe limitations on university autonomy. The proposed charter was too much even for the conservative third Duma to swallow.

Although Shvarts's successor, Kasso, who was appointed in 1910, did not try to revive the project of a new university charter, he did prove more successful in mounting a frontal assault against the bastions of the liberal faculty in Moscow, St. Petersburg, and Kiev. The immediate spark for the university crisis was the student demonstrations and protests that followed the death of Lev Tolstoi on November 7, 1910. The demonstrations led to a revival of student activism, and a concern on the part of the Prime Minister Stolypin about the ability of university administrations to keep order on campuses. On January 11, 1911, he ordered local governors and police officials to enter the universities at the first sign of trouble and break up student meetings. The same day, he instructed the rectors that they must ban all student meetings except those of a scholarly nature.[168] Since 1905 it had been the right of faculty councils to call in the police, and faculties had done so to break up illegal student *skhodki*. But now Stolypin had turned the tables; the faculty and their elected administrators were required to call in the police to break up student meetings. If they did not, local governors could do so and could arrest the offending students. With the stroke of a pen Stolypin had taken away a right to determine how best to keep order on campus which local faculties and rectors had enjoyed for nearly five years. Nonetheless, local faculty councils complied with Stolypin's decree, a compliance which sparked student strikes. When the students tried to disrupt lectures and close down the universities, the government sent bayonet-wielding police to patrol the classrooms.[169] Police were instructed to write down the names of all students they considered engaged in illegal meetings. Inevitably, the police detained some students who in fact opposed the student strike.

Such measures took student discipline out of the hands of the local university administrations and made their work next to impossible. The elected rectors and assistant rectors at St. Petersburg University and Moscow University resigned their administrative posts. Kasso, after consulting with the prime minister and the Tsar, promptly fired Manuilov, Menzbir, and Minakov, the top three administrators at Moscow University, from their professorial posts as well. Stolypin and Kasso intended this harsh measure as a warning to the rest of the faculty. "In an interview with some members of the State Council, Stolypin also repeated his determination not to allow the student strike to close any of the universities. Any professor refusing to lecture would suffer the same fate as Manuilov, Menzbir, and Minakov."[170]

On February 2, 1911, the faculty council at Moscow University met and warned that if their three colleagues were not reinstated on the faculty, many other faculty members would resign. This may have been precisely what the government desired. At any rate, Stolypin and Kasso stood fast. As a result, 28 percent of the Moscow faculty (both junior and senior faculty) resigned in protest, including Vernadsky. The resignations included 44 percent of the junior and senior faculty in the natural sciences. While one can understand the professors' dilemma, the gesture of solidarity with their fired colleagues accomplished at one swoop what the Tsarist government had desired for years: the virtual destruction of one of the institutional strongholds of the liberal opposition. Although these resignations damaged severely the academic quality of Moscow University, especially in the natural sciences, 28 percent of the faculty was a far cry from the majority support liberals had enjoyed in 1905. Only at Kiev Polytechnic Institute did as large a percentage of the faculty resign in protest over the Kasso affair. Kasso clearly emerged in firmer control of Russian higher education as a result of these resignations. Between 1911 and 1914, Kasso treated the remaining liberal faculty with disdain, transferring a number of them from the central universities in St. Petersburg and Moscow to more conservative provincial universities and importing conservative professors from the provinces to replace them.[171]

The majority of the senior faculty supported Kasso or agonized in relative silence over their own unwillingness to resign. The professors who resigned, including some of Russia's most outstanding natural scientists like Lebedev, Timiriazev, and Vernadsky, were barred from teaching in the state universities and had to find other employment.[172] The majority of their colleagues either accepted the fact that the minister of education was in charge, or they actively supported such domination as a guarantee of order on the campuses. Many of those who had taken liberal positions in 1905 no doubt agonized over their own decision to stay at their posts; but other sources of scholarly employment in Russian society were so negligible that hard economic necessity probably forced many of them to swallow their pride and retain their teaching posts. Those who resigned had to scramble to find sources of private employment, for example in one of the new private universities like Shaniavsky University in Moscow. The private Institute for Scientific Research was established in Moscow to provide laboratories for distinguished scientists such as Lebedev and Timiriazev, who had resigned from Moscow University; but the private sector simply was not strong enough to provide secure positions for everyone who had resigned. In early 1911, sixty-six important factory owners and businessmen published a statement blaming Stolypin's government for the crisis in the universities. "In the important matter of national construction," they wrote, "wrath is a poor guide to action."[173]

Of those professors who resigned in 1911, Vernadsky was one of the fortunate few who already had an alternative source of employment, having been elected an adjunct member of the Academy of Sciences and director of the Mineralogical Museum in St. Petersburg in 1906. After his resignation from Moscow

University in 1911, Vernadsky moved to the nation's capital full time and devoted himself to geochemical research and his duties as a scientific administrator.

Most of the laboratory research Vernadsky published in the field of geochemistry between 1908 and 1914 concerned the distribution of rare elements such as cesium, rubidium, scandium, and indium. In 1908 Vernadsky took issue with Professor Eberhard of Potsdam, who maintained it was impossible to observe orderliness in the distribution of scandium or in the formation of minerals containing the element. Vernadsky argued that Eberhard's own data presented a different conclusion. On the basis of the German scientist's material, Vernadsky concluded that "scandium is concentrated in those parts of the earth's core connected with outcroppings of acidic magmas and becomes isolated in bodies formed during the high-pressure stages of their congealing."[174] The professional difference of opinion between the two scientists apparently did not affect their collegiality, since Vernadsky later established close scientific ties with Eberhard in 1911. Vernadsky's correction of what he considered errors was not limited to the work of other scientists. In 1909 he published a revision of one of his own articles after further laboratory research indicated that his original thesis was incorrect. In a work published in 1908, Vernadsky had concluded that the rare element cesium is not found in feldspar. In 1909 he revised that conclusion, indicating that his lab had found cesium in some classes of feldspar.[175] During this period Vernadsky did not publish a great deal of work based on his own concrete laboratory research, but what he did publish was concerned with the distribution of rare elements. In an article on rubidium published in 1914, he put forward a generalization about the process of concentration and association of rubidium. Vernadsky maintained that rubidium always appears in association with potassium and appears in all minerals containing potassium. His lab work led him to the conclusion that rubidium plays a much greater role in the chemistry of the earth's crust than previously thought, in part because of this association with potassium. However, he considered it a mystery why more rubidium was found in ocean water than in fresh water. In ocean water, rubidium was found concentrated not only in sea organisms but in mica as well, which in Vernadsky's experience always contained that rare element. In fresh water, rubidium was much more difficult to find. Thus, his work on rubidium raised many questions, and he concluded his article with a call for further research: "In such a way we see that we are now faced, in the history of the rubidium of the earth's crust, with more questions than exact data, demanding extensive work."[176]

During this period, Vernadsky extended his interest in the paragenesis, or interactive influence in the formation process, of minerals to a study of the paragenesis of chemical elements. Following in the footsteps of several nineteenth century German and Russian chemists and mineralogists, Vernadsky's work on the paragenesis of minerals had begun back in his early days as a graduate student. Now, however, he developed an interest in mineral compounds found together in the earth's crust and in the individual atoms associated with each other as well. Vernadsky's interest in the fact that potassium and rubidium always seemed to be found together was just one example of his larger theoretical interest. While the paragenesis of minerals and atoms was not a new idea in world science, Vernadsky transferred the ideas from chemistry to the earth sciences.[177] In an article published in 1909, Vernadsky proposed that all the chemical elements can be grouped together in a limited number of isomorphic series, according to their association with other chemical elements in nature. He argued that there are only eighteen isomorphic series and that these groupings are different from Mendeleev's periodic table of elements, since they are established not by chemical composition but by their genetic qualities of association through time with other elements. "In general we know that if in the analysis of a given natural body we find some elements of a particular isomorphic series, we must seek all the others and normally we will find part of all of them."[178] Vernadsky added an important, complicating factor in defining these associations, a factor until then not recognized by other scientists.[179] "Observation," Vernadsky wrote, "tells us that isomorphic series change under the influence of temperature and pressure. A chemical element falling in one isomorphic series under a particular temperature and pressure will not enter it under another."[180] Thus, he concluded, "it is necessary to introduce a corrective, to take into consideration the thermodynamic environment in which the substance studied was formed."[181]

Here we have a good example of how Vernadsky's mind worked, jumping from very specific concrete studies, such as that on rubidium and its association in nature with potassium, to very broad theoretical ideas, such as his work on isomorphic series. The fact was that he was often bored by detailed experimental work and laboratory analysis and preferred to leave such work to his assistants. He liked to put forward large problems and major generalizations, often based on concrete data discovered by other scientists, but also in part on more limited and partial research from his own laboratory. Even though there is an unfinished quality about much of Vernadsky's work in these years, a sense of his own impatience with the boring work of laboratory analysis and a tendency to jump around from topic to topic as the spirit moved him, many of his ideas were also very stimulating and important to the development of a new field like geochemistry. In these years, for example, Vernadsky published an important scientific paper on the gases of the earth's crust and atmosphere and the role they play in geochemical cycles. In an article published by the Academy of Sciences in 1912, he stressed the great importance of natural gases produced by chemical, physical, and biological processes to the chemistry of the earth's crust. He decried that fact that geochemistry of the earth's crust was one of the most neglected topics in the earth sciences, particularly in Russia. In a letter to his wife in 1911, Vernadsky noted that he was already working on the gases of the earth's crust and indicated that "it seems I am the first to do it; at least I know of no similar work."[182] Vernadsky's 1912 article, "O gazovom obmene zemnoi kory" [On gaseous exchange of the earth's crust], was important for its emphasis on the fact that all the earth's gases, as well as earth's minerals, appear to be involved in a series of cyclical processes, closed cycles that continually repeat themselves to maintain a constant chemical process over long periods of time. Citing the research of other scientists, Vernadsky indicated that science already knew a good deal about such things as the oxygen cycle and the nitrogen cycle, but indicated that little was known about the history of methane or of helium and other noble gases. This was particularly true for the natural gases found within the borders of the Russian Empire. "At every step we stand before puzzles. For Russian scientists here is a huge field of work, but the gases arising from the earth's depths in the borders of Russia are completely unstudied. . . . Natural gas is a mighty source of energy and this energy either is not touched by us in Russia or is stupidly expended for no purpose and without use. It can be intelligently used only when it has been studied scientifically."[183] Once again, Vernadsky set a new research agenda for Russian scientists in an area previously neglected, the problems he defined for future work being more interesting than the rather tentative conclusions he had reached.

In 1909, Vernadsky first read F. W. Clarke's book, *Geochemistry*, which had appeared the year before and gave his reaction to his former student, Samoilov, now a professor of mineralogy at the agricultural institute in Moscow: "I am reading Clarke's *Geochemistry*, a *very interesting* and good book. Many things have been left out, of course (for example about organic matter, rare elements and their dispersion, the living life of the sea, nitrogen-fixing bacteria, etc.) but there is much for me that is new. . . ."[184] Vernadsky was particularly intrigued by the role played by living matter in the geochemistry of the earth's surface and atmosphere, a subject not explored by either Clarke or Goldschmidt. In fact at least as early as September 1906, Vernadsky wrote in his unpublished notebooks: "What significance has the entire organic world, taken as a whole in the general scheme of the chemical reactions of the earth? Has the character of its influence changed in the course of all geological history and in what directions?"[185] Vernadsky wondered whether organisms inevitably played a role in all geochemical cycles and asked himself what that role was.[186] He began to read widely in biological literature as well as works in physics, chemistry, and geology. In 1908, he wrote his son:

> My thought is occupied with a new area which I am embracing—about the quantity of living matter and about the interrelationship between living and inert matter. With some awe and lack of understanding, I am all the same entering this new area, since it seems that I see some sides of a problem which until now no one has seen. I am succeeding here in approaching new phenomena.[187]

To be specific, Vernadsky turned a geologist's mind, trained among other things to think in terms of masses, strata, and formations of matter, on biological phenomena. Whereas biologists for the most part were still studying separate living beings, their communities and species, and their evolution through time, Vernadsky was beginning to think of all life almost as if it were a geological stratum, a single mass which he called "living matter"—a type of matter that was highly chemically active and different in many respects from non-living matter. Vernadsky began to ask how much living matter existed, as a whole, and what its role was in the cycles of various chemical elements. While "living matter" evolves through a multitude of different, non-repeatable forms, Vernadsky was quick to recognize that at the level of atoms chemical elements go through closed cycles that continually repeat themselves.

While it seems an exaggeration to agree with one of Vernadsky's biographers that he was the "Darwin" of mineralogy, Vernadsky was one of the first scientists to ask what role "living matter" played in these geochemical cycles and how living matter and inert matter differ.[188] He also came to the conviction that although the forms of living matter evolve and are not repeatable—species like the dinosaurs, for example, become extinct and do not reemerge in endless cycles—perhaps underneath the evolutionary process with its endless changes of form, the overall quantity of living matter on earth remains a constant. Only the forms change and evolve.

I have been unable to discover where Vernadsky acquired this conviction and on what evidentiary basis. He was widely read in both science and philosophy, so it is difficult to say. He had also long had a mystical bent which he generally suppressed in his published work but which comes out in his unpublished correspondence and notebooks and even occasionally in print. This was a bent, already mentioned with reference to his youth, which drew him to Eastern religions and philosophy, the speciality of his friend, Sergei Oldenburg, the permanent secretary of the Academy of Sciences. As Vernadsky put it in one of the letters to his wife in this period after the 1905 revolution: "It seems to me that in my work now, I am proceeding partly unconsciously. I have a strange feeling as if some kind of unknown force is leading me."[189] We have another hint of his attitude toward mysticism in the tribute he gave in 1908 to his friend Prince Sergei N. Trubetskoi, the deceased rector of Moscow University, who was a religious mystic and critical idealist philosopher. In this tribute, Vernadsky indicated his belief that "mysticism is one of the very deepest sides of human life," given the faith it promotes in the future and its belief that unknown forces are guiding humanity in purposeful directions that give ultimate meaning to life. Vernadsky, who had a very strong belief in intuition as a source of scientific creativity, felt that mysticism was not necessarily an enemy of science, if combined with knowledge and critical intelligence, as he believed it had been in the personality of Trubetskoi. In fact, he believed that mysticism, by giving a sense of faith and meaning to human life, could be an ally of science and a source of scientific intuition.[190]

Wherever his ideas on "living matter" came from, they first appeared in his notebooks and private correspondence with his son George and his former student Samoilov in the years between 1906 and 1908. In this period, such ideas appear in the form of speculation, but in his later work after 1917, they acquire more and more importance as he attempted to verify such speculations. Vernadsky's speculation on the constant quantity of living matter was closely tied in with another bit of speculation shared by some other writers of the time, the view that perhaps life had not originated on earth but had come to earth from another part of the cosmos in nucleus form. Vernadsky carried this philosophical speculation even further, wondering whether life was not only a constant in the cosmos but whether it was eternal, had always existed, alongside matter and energy. These ideas remained private speculations for Vernadsky until after the 1917 revolutions, when he began to speculate about them in print and to arouse rebuttals from other Russian scientists.

In 1908, he shared these ideas in a rather disjointed form with his former student I. V. Samoilov:

> I have thought much recently about the question of the quantity of living matter, about which I spoke to you earlier. I am reading in the biological sciences. [The concept of] mass for me is curious. The conclusions reached force me to think. By the way, it turns out that the quantity of living matter in the earth's crust is a constant. Then *life* is the same kind of part of the cosmos as energy and matter. In essence, don't all the speculations about the arrival of "germs" [of life] from other heavenly bodies have basically the same assumption as [the idea of] the eternity of life?[191]

Although the logic of these rather cryptic remarks is not very clear, they do provide evidence of when Vernadsky began to think about the role of life on earth and in the cosmos and what kinds of questions he was posing for himself.

Since Vernadsky presents no proof or source of these ideas, and I have been unable to find contemporary sources in the scientific literature that he might have derived them from, I have reached the tentative conclusion that here, as in some other parts of Vernadsky's work, he simply formed hypotheses that went well beyond any scientific evidence he had. Such hypotheses may have been derived to some extent from a mystical strain in his thinking, but this is perhaps not as important as the fact that these hypotheses helped to guide much of his future scientific research program. It is probably no coincidence that his interest in meteorites, which stemmed in part from his search for evidence of "living matter" and its products in other parts of the cosmos, and his development of the concept of the biosphere and its connections with the rest of the cosmos date from precisely the same years, 1908–14.

Vernadsky was not the first scientist to use the term "biosphere," which appears in his published work for the first time in 1914. It had been used earlier by the Austrian geologist, Eduard Suess, a contemporary of Vernadsky and someone whose works Vernadsky not only knew, but with whom he had a personal acquaintance. Vernadsky had visited Suess in Vienna in 1910 when he was lining up cooperation from West European scientists for his project of creating a map of all known deposits of radioactive minerals on earth.[192] Suess had used the term biosphere in passing, and Vernadsky wished to give it a precise quantitative and qualitative meaning by studying the geochemical cycles occuring there and discovering the role life plays in these cycles. The biosphere, which he defined as that part of the atmosphere and surface of the earth where life exists, (he later gave it a much more precise geographical definition), was of greatest interest to him because it was the most chemically active part of the earth and because he believed that virtually all of the minerals formed in the biosphere were formed in connection with biological processes. "At the very surface of the earth—in the region of the biosphere—it is scarcely possible to speak of the preservation of chemically unchanged matter of any kind over the course of millions of years." Vernadsky believed that this was due in large part to the chemical role played by "living matter."[193]

In addition to concrete laboratory research and broad theoretical statements, Vernadsky in the years between 1908 and 1914 carried out a series of expeditions to the more remote parts of the Russian Empire in order to map the known deposits of radioactive minerals there and, he hoped, to find new deposits. As the scientist noted in one of his articles during this period, no new deposits of radioactive minerals had been found recently, and the shortage and high cost of such minerals was holding up research and their practical utilization.[194] Vernadsky felt the situation to be particularly urgent as medicine had discovered the therapeutic value of radioactivity. During one of his expeditions to Central Asia in 1911, Vernadsky wrote to his wife telling her he wanted to be well versed on the mineralogy of Russia, not only its scientific significance, but its practical side as well. One of his goals was to establish contact with as many miners as possible, in the hope that they would send samples of unusual minerals to the Academy of Sciences for analysis.[195] Between 1908 and 1914, Vernadsky sent at least one expedition yearly to the more remote and promising areas of the Russian Empire in search of radioactive minerals. Expeditions were led to the Caucasus, Central Asia, the Urals, and Siberia. Vernadsky led some of these expeditions personally; others were led by his students. By 1914, he had as many as thirty persons engaged in such expeditions, which were often dangerous and sometimes disappointing in their results. In an expedition to the Caucasus in 1911, Vernadsky nearly drowned in a swift mountain stream, and several times almost slipped from steep mountain paths, all without finding a single trace of uranium, one of his major goals on this expedition.[196] He had somewhat better luck the same year in Central Asia, where signs of uranium were found, although the most promising was in a rundown mine whose owner was absent. Vernadsky and his expedition were afraid to go to the bottom of the mine because of its poor condition, but he noted that the Fergana region of Central Asia seemed promising. The Fergana was a remote area of camel caravans, Kazakhs who spoke no Russian, strange birds which watched them as they stared back, and violent windstorms where sand penetrated all their supplies and made their lives miserable.[197] With the outbreak of World War I,

Vernadsky was unable to return to this area but many decades later he remembered this expedition to the Fergana region and suggested to the Soviet Academy of Sciences that it launch a major effort to find uranium deposits there as it geared up for its own nuclear energy program during World War II.[198] In 1912, one of Vernadsky's students, K. F. Egorov, found an even more promising uranium deposit near Lake Baikal which he named after Vernadsky and which also later proved of great value to the Soviets.[199]

But it was in the Urals in the years before the first world war that Vernadsky and his helpers found their richest variety of radioactive minerals, in a remote area around Lake Ilmen. The area had been mined off and on for more than two hundred years, but often quite wastefully. Vernadsky remarked in 1911: "The Urals produce a heavy impression by the terrible plundering of its riches . . . forests, mines of precious stones, roads, forms of life—all reflect the disorder of this antediluvian governmental structure and the anarchy which reigns everywhere! You cannot imagine what barbarism is found in the famous Murzinka region and its vicinity! Yet, the wealth here is enormous. In 200 years there is not yet one respectable road! The forests burn and are wasted. In order to obtain precious stones, almost half are destroyed and future work is made almost impossible."[200]

This outburst marked the beginning of Vernadsky's campaign, first with the Tsarist government and then with the young Soviet government after 1917 to put the Lake Ilmen area under state protection. The campaign succeeded in 1920, when a decree signed by Lenin made the area the first nature preserve or national park in Soviet Russia. The 1920 decree was largely the result of agitation by Vernadsky and several of his more prominent students.[201]

On his expedition to the Urals in 1911, Vernadsky met two of his most valued assistants, L. A. Kulik, a forester with encyclopedic knowledge of the minerals of the Ilmen region, and N. M. Fedorovsky, who later became an important official of the Bolshevik government and helped rescue Vernadsky from a Petrograd prison in 1921. Kulik was not only of great help to Vernadsky in exploring the Urals region, he was later, at Vernadsky's urging, to become the world's greatest expert on the Tungus region of Siberia. A large portion of this Siberian wilderness had been devastated in 1908 by a cosmic fireball of unknown origin, an event which had fascinated Vernadsky at the time. Vernadsky later sponsored the creation of a Commission on Meteorites in the Academy of Sciences after the revolution; and in 1927, he helped Kulik form an expedition under its sponsorship to study this mysterious event in the Tungus region.[202]

In the summer of 1911, Vernadsky made a different kind of addition to his school in the person of N. M. Fedorovsky, an expelled student with a radical political backgound. Fedorovsky's mother had been a school teacher and a member of the People's Will organization, while Fedorovsky himself had been an active Bolshevik as early as 1904 and had participated in the revolution of 1905, urging sailors of the Baltic fleet to rebel. Fedorovsky had entered Moscow University in the fall of 1908, choosing to major in physics and math. During the stormy events of 1911, he had participated in an illegal student *skhodka* and had been expelled. Left entirely without money, he tried to sell his rock collection to the Moscow firm of Ilin, which in addition to maps, sold rare minerals. Instead, the firm proposed that he go to the Urals and collect minerals for them. Although he had not yet met Professor Vernadsky, Fedorovsky sought advice from Vernadsky's assistant, Aleksandr Fersman, who at the time was the curator of the mineralogical collection at Moscow University. Fersman recommended that he go to the Ilmen mountains in the Urals if he wished to find valuable minerals. Financed by the Ilin Company, Fedorovsky did so, where he met Vernadsky later in the summer of 1911.

Vernadsky drove up to the empty schoolhouse in the Urals town of Miass where Fedorovsky was packing the minerals he had found. Impressed by Fedorovsky's collection and his knowledge of minerals, he invited the young man to join his expedition. Learning that Fedorovsky had been expelled from the university, he also promised to help him be readmitted, with two conditions: Fedorovsky should change his major to mineralogy and he should sever his ties with the Ilin firm, since Vernadsky did not approve of the commercial exploitation of this area. Fedorovsky agreed to the terms and decided to make the study of minerals useful to Russia his life's work. He was soon readmitted to Moscow University, where he received his degree in 1914, the same year he published his first scientific work, *Granites in Nature and Technology*. Despite the fact that Fedorovsky remained active in the Bolshevik underground in Moscow during these years, he was invited to remain at Moscow University and prepare for a career as a professor. His career, however, was interrupted by the outbreak of World War I and the 1917 revolutions, in which Fedorovsky was to play an important part as a leader of the Bolshevik party in Nizhnyi Novgorod during the October revolution. Following the civil war, Fedorovsky became an important industrial administrator in the Supreme Council of the National Economy (VSNKh), where he was in charge of the mining industry. Fedorovsky eventually became head of the Mineralogical Institute of the Academy of Sciences, which he administered until his arrest and imprisonment during the Stalinist purges of 1937.[203] Fedorovsky always considered himself a student of Vernadsky, despite the fact that Vernadsky himself moved to St. Petersburg in 1911 and Fedorovsky worked most directly with Vernadsky's students who remained in Moscow.

Radical students come to know Vernadsky in these years for his kindness toward them. Their common opposition to the Tsarist government brought them together, despite frequent differences in political and philosophical outlook. In addition to helping Fedorovsky's career in mineralogy, in 1912 Vernadsky volunteered to testify at the trial of one of his former students in the Moscow Higher Women's Courses, A. B. Missuna. Missuna had been arrested for "radical propaganda" during a summer geological expedition. Although she was not immediately released, Vernadsky's testimony as a member of the Academy of Sciences helped her receive a reduced sentence.[204]

In general, Vernadsky was appalled by the waste of human talent in Russia, which he blamed on the Tsarist government. In the summer of 1913, he traveled to Canada to attend an International Geological Congress and then to the United States. He was struck by the number of Russians living and working productively in Canada and was depressed by the fact that their talent was lost to Russia, thanks to persecution by the Tsarist government or lack of economic opportunity.[205] Russia's scientific backwardness was brought home to him vividly when he visited the United States, especially the scientific institutions of Washington, D.C., such as the Carnegie Institution. At the time, Vernadsky had a proposal before the Imperial Russian government for the creation of a similar institution, affiliated with the Academy of Sciences, to be called the Lomonosov Institute. The proposal had not yet been accepted. In a letter to his wife, he noted that:

> Over the past ten years the United States has made huge advances in science: today America gets along without the help of German universities, which not long ago was considered indispensable. When I involuntarily compare these years in Russia—the activity of Kasso and Company—that whole gang that constitutes our Imperial government—I feel depressed and uneasy.[206]

Vernadsky felt that Russian science and creative thought had advanced in this period, in spite of the Russian government rather than because of it. The comparison with the new world was a depressing and unsettling one. American scientists, Vernadsky noted, complained that business and businessmen were valued more highly than science and scientists in America, but all that is "nothing compared with the conditions under which [we] have to work."[207]

Vernadsky was more impressed by the quality of North American science than he was by the effects of technology. On visiting the nickel and cobalt mines at Sudbury, Ontario, he wrote:

> This new technology—American technology—which has given so much to mankind, has its dark side. Here we see it in everything: a beautiful land has been made ugly, the forest burned out; for tens of miles the land turned into a wasteland, all plant life poisoned and burned out, and all of this in order to achieve a single goal: the quick mining of nickel.[208]

Vernadsky's environmental awareness was sharpened by what he saw in North America, and he felt himself more nationalistic and Russian than ever, determined that such devastation of the natural riches of Russia for profit be avoided as his nation industrialized, an attitude which after the 1917 revolutions led to the creation by Lenin and the Soviet government of the first Russian nature preserve, at the urging of Vernadsky and Fersman.

The year 1913 was memorable not only for Vernadsky's first and only trip to North America, but also for the fact that he turned fifty that year. Personally, Vernadsky was prospering, both materially and professionally, despite his loss of the professorship at Moscow University in 1911. He had numerous devoted students; he was a full member of the Imperial Russian Academy of Sciences, where he headed the Geological Museum and enjoyed the use of a spacious eight-room apartment. He was Russia's leading expert on radioactive minerals and one of the world's leaders in the burgeoning field of geochemistry.

Vernadsky's personal property was also increasing, although he had never been particularly materialistic, a sign of increasing security on his part. He was building an eleven-room dacha on a large lot in the Ukraine at Shishaki. Located above a bend in the Psel river, the dacha was covered by an oak forest and at night had a view of the lights at Sorochinsk. Vernadsky spent many pleasant summers there until 1917, working on his writings, while surrounded by his family and numerous friends who came to visit. In addition, he purchased a lot near the Black Sea in the Crimea where many other academic friends were buying in order to escape the unhealthy climate of St. Petersburg. His income from his estate at Vernadovka had improved since 1912. He had fired the steward of his estate, Popov, because some of Vernadsky's friends had convinced him that Popov was cheating him. Vernadsky's son, George, disapproved of his father's action, and Vernadsky himself apparently later regretted his decision, since Popov had worked loyally for him for many years.[209] Nonetheless, in 1912 Vernadsky rented Vernadovka to the Dolgorukov Sugar Company, retaining for himself and family only the house and a small garden plot and thereby increasing his income.[210]

More important to Vernadsky than personal wealth was the love and affection of his family and friends. His wife was utterly devoted to his well being. As one of Vernadsky's students, L. V. Vasileva, whom he hired for two years in this period as a tutor for his daughter, described her:

> Nataliia Egorovna—his friend and wife—was a humble, highly cultured, kind and sensitive woman. She was a good housewife, devoting her whole life to caring for her children and her husband. She worked hard to preserve his health and peace and in this way aided his brilliant creativity.[211]

Vasileva idolized Vernadsky as a teacher and friend. When he lectured, she wrote, "the rocks talked."[212] In 1906 or 1907, one of Vasileva's sisters, a village school teacher was arrested for supposedly stirring up the local peasants. After five months imprisonment, she was ill and in no condition for the Siberian exile to which she had been sentenced. When Vernadsky learned about this, he wrote his friend S. F. Oldenburg, the secretary of the Academy of Sciences in St. Petersburg. Oldenburg used his ties in the government to get her sentence reduced to three years exile in a provincial town under surveillance.

Not everyone, however, idolized Vernadsky, but those who did often did so excessively. Vernadsky's niece, Anna S. Krylenko, a twenty-five-year old suffering from tuberculosis, who came to live with the Vernadskys following his sister's death in 1910, felt especially close to him. She often played the harp for him on summer evenings at Shishaki prior to her death in 1917. As she put it in a letter to Vernadsky which she wrote in 1911:

> You amazed me in many respects, especially your harmonious personality: an excellent scientist and a man with a sensitive social conscience, a specialist and a broad philosopher. You are not only a visionary but a practical man who gets things done and believed in people.[213]

Among those who did not idolize Vernadsky was his housekeeper, Praskoviia Kirillovna Kazakova, who joined the family in 1909. Although her observations were tempered with affection, as she put it in her manuscript memoirs, "As a man of the world he was inept. He didn't know the simplest things." Once at Vernadovka, she observed, as he received the sheaves from the local peasants, he sat in a rain coat and hat and saw nothing: neither how many sheaves were brought nor what kind. "He sat like a real professor," she said, implying that he was easily fooled in worldly matters.[214]

Another critical view came from one of his students in this same period, E. E. Flint, who noted that Vernadsky's lectures were full of good material, but hard to follow for beginning students. He would start out systematically, but then get sidetracked on some other theme connected with the material of the lecture and it was difficult to follow his train of thought. "For us, his ideas and theories were little understandable."[215]

On the eve of his fiftieth birthday, Vernadsky himself made the harshest assessment of his life and achievements: "I am fifty years old, but it seems to me I am far from achieving in my scientific development my limit not only of a knowledge of but an understanding of the natural environment." Vernadsky felt he had failed to achieve many of the dreams and plans of his youth, but he was full of hope that he was at the beginning of a period of great creativity. And indeed he was, as the years of World War I and the civil war were to prove.

Vladimir Vernadsky's parents, Anna Petrovna and Ivan Vasilevich Vernadsky, after their marriage in 1862. (All photos are from the collection of the Vernadsky Office-Museum in Moscow.)

Vladimir Vernadsky, 1868.

Vladimir with sisters Ekaterina and Olga.

Vernadsky with fellow students
A. N. Krasnov and E. Remizov
after passing D. I. Mendeleev's
examination in chemistry at St.
Petersburg University, 1882.

Nataliia Egorovna Staritskaia,
1884. Staritskaia married
Vladimir Vernadsky in 1886.
Their marriage lasted for fifty-
six years, until her death in
1942.

Nataliia and Vladimir Vernadsky, 1910.

Vernadsky in his apartment in Petrograd, 1921.

Vernadsky and his daughter Nina, 1925.

Vernadsky with A. E. Fersman, 1939.

The Vernadskys lived on the second floor of this house on Durnovskii pereulok in Moscow from 1935 to 1945.

Nataliia and Vladimir Vernadsky, 1940.

Vernadsky in 1944, shortly before his death.

Vernadsky's grave in Novodevichy Cemetery, Moscow. Shown here in 1953 are E. M. Vilensky, the sculptor of the monument, and V. S Neapolitanskaia, curator of the Vernadsky Office-Museum in Moscow.

The V. I. Vernadsky Institute for Geochemistry and Analytic Chemistry, USSR Academy of Sciences, Moscow.

CHAPTER

4

SCIENCE, WAR, AND REVOLUTION
Vernadsky and Soviet Science in Transition, 1914–1922

When World War I broke out, Vernadsky was leading an expedition to look for radioactive minerals in a remote area of Siberia, beyond Lake Baikal, near the Manchurian border. When word finally reached him, he hurried back to St. Petersburg. As he wrote his friend and former student Samoilov, "I landed here [at Chita] in the middle of mobilization, and then this catastrophic war with Germany broke out. What was prepared after 1871, more than forty years ago, has taken place. . . . Who will win? Whoever will, we are at the center of one of the world's most profound events."[1] Vernadsky saw war as a danger in that it threatened the integrity of the Empire because Russia was so ill-prepared to enter a war with Germany, which supplied many of its industrial needs and strategic minerals. For example, shortages of metallic bismuth for pharmaceuticals and of tungsten for shrapnel, both previously supplied by Imperial Germany, quickly occurred.[2] But Vernadsky and his associates also saw the war as an opportunity to press the dynasty for constitutional reforms and to prove to the regime the indispensability of science to the war effort and to Russia's long-range safety and well-being. Vernadsky worked actively for the first goal by participation in the Central Committee of the Kadet party, which during the war met frequently in the summers at his dacha at Shishaki in the Ukraine.[3]

To meet the second goal, he published several articles in 1915 on science and the war. In one of these he noted that World War I was already the most science-based war in human history and therefore the most devastating. Although the war had stimulated the growth of science, it had been mainly in certain destructive directions. One of the tasks for science in the future, he felt, was to stimulate science to develop defensive measures to lessen the destructive effects of war. This task, he felt was no more utopian than other measures tried in the past. He was not optimistic that future wars could be avoided, but he felt that science could play a positive role in the deterrence of war and making it less destructive. Whereas science often appears to be morally neutral, Vernadsky felt that scientists do not have to be. They can actively work for employing science in more positive ways. Right now, he felt the effects of war on science were primarily harmful. Not only does war drain material resources away from science, it has caused the death or maiming of thousands of scientifically talented persons. War also tends to undermine the idea of the unity and commonality of all human beings, which science, as an international institution, reinforces. The growth of science depends on the rapid dissemination of the results of research, in which the Germans had played so important a role with their journals, conferences, etc., before the war. World War I, Vernadsky feared, would create such strong suspicions and hatreds among different national groups of scientists, causing great harm to science. Another effect the war would have on science, Vernadsky believed, would be to accelerate the shift of scientific creativity from the European center to the peripheries of North America and the East, a shift already observable before the war. Yet science would be needed in Europe even more after the war to overcome its destructive effects. So increased attention to and appropriations for science would be necessary. In Russia, he argued, the chief task would be to learn what the country's natural resources were and to lessen that country's dependence on Germany. For this purpose, scientists, with much more government aid than before, would need to organize a whole network of labs, museums, and institutes. This task, he asserted, was not any less important than improving the general condition of Russia's civil and political life.[4]

To follow through on some of these ideas, in 1915 Vernadsky became the moving force behind the creation of a commission within the Academy of Sciences, the Commission for the Study of the Natural Productive Forces of Russia (referred to hereafter as KEPS). In January 1915, he wrote the memorandum proposing such a body, which would bring together existing knowledge of strategic war materials and gather new information about Russia's resources. Endorsed by the physical-mathematical section of the Academy and such leading Academicians as Karpinsky, Golitsyn, Kurnakov, and Andrusov, Vernadsky was elected its chairman and his student Fersman was chosen as its secretary and chief administrator. Although the commission did a great deal of valuable work in gathering information and publishing it, the immediate goal of materially aiding the war effort was, by and large, not achieved. As Vernadsky indicated in a letter of August 1915 to Fersman: "The situation has become so serious that for the near future we in our commission perhaps cannot do much; but the long-range task has become even more important and demanding, since the war will be longer than I thought and our losses will be greater. . . ."[5]

A series of roadblocks impeded the activity of Vernadsky's commission. The greatest was a lack of funding and support by the Tsarist regime, which mistrusted the motives and activity of politically active liberals like Vernadsky, preferring to work through its own government bureaus. Fersman recalled that "even in solving the most vital problems, such as developing deposits of tungsten, the Academy of Sciences in the course of two years was unable to obtain financing. Specialists in the study of the Caucasus confirmed a series of interesting deposits of sheelite on the northern side of the Caucasus. But the ministry refused an appropriation of 500 rubles to confirm these deposits. I recall that at one stormy session [of KEPS], after the refusal of the Tsarist government to grant credits was announced, one of the members of the commission, Academician A. N. Krylov, with a passion unique to him, declared that this chaos must be put to an end."[6] Krylov's anger was sharpened by the fact that a rich deposit of tungsten was known to exist in Turkestan, but the commission was not permitted to study it because it was located on the lands of one of the Grand Dukes. Krylov commented bitterly that these lands should be requisitioned because if the war was lost, not only would the Grand Dukes lose everything, but the whole dynasty would go to the devil's mother.[7]

Despite these obstacles, the activity of the commission was not completely fruitless. In 1916 alone KEPS organized fourteen special expeditions to look for strategic minerals.[8] In 1915 Fersman and another scientist announced the existence of possible bauxite deposits in Russia, and on March 18, 1915, Vernadsky and Fersman sent a memorandum to the Ministry of Trade and Industry, proposing the creation of an aluminum industry in Russia, a venture that was not realized until after the revolution.[9] In December 1916, Vernadsky also proposed a series of practically oriented research institutes within the Academy of Sciences to deal with economic problems. His recommendation was ignored by an already disintegrating government.

Fersman later characterized the overall activity of KEPS during the war: "Considering the economic backwardness of Tsarist Russia, the practical effect of the activity of KEPS was insignificant. Its work in the main consisted of making inventories of resources, studying known deposits, and drawing up plans and proposals." Although written by Fersman during the Soviet period and obviously meant to put the Tsarist regime in the worst light by comparison with the Soviet, this does seem to be an accurate assessment.[10]

When he was not preoccupied with the work of KEPS, Vernadsky spent three to five hours a day studying radioactive minerals, such as samples of uranium from Colorado sent to him by a scientist he had met in Washington, D.C. By 1916 Vernadsky was more and more concerned about the political situation. Although he was optimistic about Russia's long-term future, he was worried by the growing mood of anger among most groups in Russia, especially at the conduct of the war and the influence of Rasputin on the autocracy. He wrote to Aleksandra Vasilevna Golshtein while sitting through some of the boring speeches of the State Council, to which he had once again been elected in 1916: "We are living through a terrible time here." Nonetheless he was optimistic: "Perhaps because of my nature, perhaps because I live among people who are working for the positive construction of Russian life, I look calmly at the future of Russia and am confident that in the end we will emerge from this world conflagration intact."[11]

One of the themes that was to emerge from much of his correspondence over the next several years was his confidence in the Russian people and the need for educated Russians to preserve and extend their cultural work to bring the Russian people to a higher level. When his son, George, finished his advanced degree in history at Moscow University and decided in 1917 to join the army, from which he had been exempted while a student, Vladimir advised his son not to join as a volunteer: "Teachers and scholars are exempted by the government; in Russia this is a good policy, since there are so few scholars. But if your feelings lead you to do this, then I understand. Reason is often not the most powerful or important force in life. But . . . Russia needs teachers. It is better to help the war effort as a civilian, behind the lines."[12]

After the collapse of the Tsarist regime in early 1917, Vernadsky became even more involved with organizational activity than before. There were so few people with his talent and willingness to serve that he was asked to participate in a series of new committees, including an agricultural science committee composed of some of the most distinguished scientists in Russia. The aim of this committee was to increase the productivity of agriculture through the application of the latest scientific techniques. Vernadsky was chosen chair of this body, as he was of so many other groups. In March 1917, he was also elected head of the Department of Mineralogy and Geology at Moscow University, the position he had held until his resignation in 1911. Once again he faced the prospect of commuting between St. Petersburg and Moscow. And finally, in early 1917, he began to formulate the idea of a science city, which would unite in one location people in many fields of knowledge. He had even picked the site: the Palace of the Emperor Paul I at Gatchina, about an hour's ride from Petrograd. The palace had more than 600 rooms and was surrounded by a 1,500-acre park, which could also be used for scientific purposes. Although his idea was not realized at this time, the concept of such science cities is significant for the history of Soviet science; the idea was finally realized after the death of Stalin.[13]

Vernadsky was spreading himself and his energies more and more thinly. No longer a young man, his varied duties began to take a toll on his health. In the spring of 1917, he was told by his physician that he was suffering from tuberculosis, the disease that his niece, Anna Sergeevna Korolenko who lived with the Vernadskys, died from in 1917. He was advised to leave the unhealthy climate of Petrograd and travel to the south as soon as possible. However, given his conscientiousness, he felt unable to leave until June. After a short stay in the Ukraine taking a kumiss cure (fermented mare's milk), he was back in Petrograd in July. His friend, S. F. Oldenburg, had been appointed minister of education in the Kerensky government and asked Vernadsky to become assistant minister in charge of all universities and scientific institutions in Russia. In his short tenure as a government official, Vernadsky formed committees to plan for the reform of higher education and science in Russia, but this was not a time for the slow, deliberate, liberal way of doing things. Nonetheless, a new university was created in Perm during his tenure as assistant minister largely because of his belief that higher educational and scientific institutions in Russia were too highly concentrated in Petrograd and Moscow. Discussions also were held about the creation of three new academies of science, in the Ukraine, in Georgia, and in Siberia. But as Vernadsky recorded in his diary, the popular mood did not support the provisional government and time was short. Three days before the Bolshevik seizure of power, he wrote in his diary that "in essence, the masses are for the Bolsheviks." The day before the October revolution he confided: "The future seems murky and threatening. All the same I feel very clearly the strength of the Russian nation."[14]

After the October revolution, Oldenburg and Vernadsky turned over their portfolios to V. I. Lunacharsky, the new Bolshevik commissar of enlightenment. Lunacharsky did not invite them to remain in the commissariat. Oldenburg stayed in Petrograd as chief administrator of the Academy of Sciences, where he maintained an uneasy relationship with the Bolsheviks over the next few years. Vernadsky headed south for the Ukraine, where his family was staying with relatives, intending to concentrate as much as possible on his scientific work.[15]

He initially settled in Poltava, where his family was staying with his brother-in-law, but soon a letter came from N. P. Vasilenko, a Ukrainian historian in Kiev with whom Vernadsky had held talks while he was assistant minister of education in 1917. Vasilenko, on behalf of other Ukrainian intellectuals, wanted to revive the idea of creating a Ukrainian Academy of Sciences, and in May 1918, they invited Vernadsky to move to Kiev to help them. Feeling sufficiently recovered from his bout with tuberculosis, Vernadsky was eager for active involvement in organizational work. He remembered later that he

> quickly came to an agreement with Vasilenko and was very attracted by the possibility of creating an academy. I set a condition that I should take part in the cultural work of the Ukraine in the capacity of an Academician of the Russian Academy of Sciences, that is, in the role of a consultant. . . . It seems to me that I was then the only person in Kiev who was familiar with the work of an Academy in a practical sense, as it then was organized in the Petrograd Academy of Sciences.[16]

Vernadsky soon moved his family to Kiev, where they found quarters for a time in an abandoned high school, living out of cardboard boxes and orange crates. More than a dozen governments came and went in the course of 1918–20, yet despite the chaotic conditions in Kiev, Vernadsky continued his scientific research, was elected President of the Ukrainian Academy of Sciences in 1919, created a Ukrainian KEPS, and participated actively in the Geological Committee of the Ukraine. As his housekeeper, Praskoviia Kirillovna, later put it: "Wherever he went, Vladimir Ivanovich set up his own little shop."[17] He was eager, as he wrote Fersman in Petrograd, for the Ukraine to maintain close ties with Russia and felt himself caught between two enemies: "extremist Ukrainians" and "extremist Russians."[18] Fortunately, Vernadsky was able to maneuver between the two groups. On the one hand, he spoke Ukrainian and was interested in the development of the Ukrainian nationality and its culture. At the same time, he did not want to eradicate Russian language and culture in the

Ukraine. He made clear his loyalty to Russia and the need for unity and cooperation between Russia and the Ukraine on the basis of equality.

Over the next months, he worked hard to raise money for the Ukrainian Academy from the many governments which ruled Kiev and parts of the Ukraine during 1919 and 1920.[19] One of Vernadsky's associates in this period, the engineering professor S. P. Timoshenko (he later emigrated to the United States and taught for many years at Stanford), recalled that the Bolshevik government was more forthcoming with aid for the new academy than either the Petliura government or the Volunteer Army.[20] However, Vernadsky's own memoir about these years indicates that after travelling to the headquarters of the White government in Rostov and talking with its officials, they recognized the need for the Ukrainian Academy and gave it financial support.[21] How much money the academy actually received from the Whites and their Volunteer Army is uncertain, since the lines of communication were frequently cut during the civil war in the Ukraine. The academy lived largely on whatever it could get from any of the existing governments.

The memoirs of Vernadsky and Timoshenko agree on one important point: the alienation they felt from all sides fighting in the civil war. As much as both men disliked the Bolsheviks, they felt little attraction toward the Whites. Vernadsky wrote in his unpublished diary on December 29, 1919: "Both the Volunteer Army and the Bolsheviks did a mass of unclean deeds; and in the final analysis one was not better than the other."[22] Vernadsky argued with other scientists in the Ukrainian Academy about whether the Bolsheviks would last or were a passing phenomenon, taking the view that the popularity of Bolshevism was fading and that their government would not last.[23] However, he wavered back and forth on this question, noting at another point in his diary that if the Bolsheviks would stop their use of terror, he thought the population would support them in large numbers.[24] A month earlier, he had already noted a loss of support for the White Army, which he criticized in his diary for being too selfish and class oriented, not thinking about the good of Russia as a whole.[25] Probably for the same reasons, and particularly because he considered most of the Ukrainian Kadets to be Russian chauvinists, he resigned from the Kadet party in this period, although publicly he gave a different reason for his resignation. He declared that membership in any political party was incompatible with his role as President of the Ukrainian Academy. By 1920, he was clearly fed up with politics. He confided to his diary that it was necessary for him to stay away from politics and concentrate on his new scientific ideas, for which he felt great excitement.[26]

Like some other prominent scholars, he associated himself primarily with cultural and educational activity rather than with active support of any of the groups contending for power. However, Vernadsky himself noted in his diary that he personally felt safer under the Volunteer Army than under the Bolsheviks or Petliura's Ukrainian nationalist regime, because of his connections with the former Tsarist government and because his high position in the provisional government made him suspect in the eyes of these other groups.[27] At the same time, he noted that most scholars of Jewish background felt safer under the Bolsheviks because of the anti-Semitism of the other groups (which Vernadsky did not share). Anti-Semitism had reached a fever pitch in the Ukraine during this period, Vernadsky believed, because of the prominence of so many Jewish commissars in the Bolshevik party. The Orthodox churches in Kiev were crowded as never before, a return to religion of which Vernadsky did not approve. In 1920, he noted that his son, George, under the influence of the Orthodox theologian S. N. Bulgakov, and George's wife, Nina, had gone deeper into Orthodoxy, and for him this renewed faith meant a revival of the Russian monarchy. Vladimir Vernadsky clearly disapproved.[28] Vernadsky considered himself neither Orthodox nor Christian, although he preferred Christians to socialists, whom he felt had a strong tendency to suppress ideas with which they disagreed.[29]

What particularly concerned scholars such as Vernadsky who had become established during the tsarist period was the indiscriminate terror often practiced at the grass-roots level against all middle-class elements by Bolsheviks in the Ukraine, as well as by the peasant-oriented Ukrainian national government of Petliura. Added to that was the rise in criminal activity and sheer anarchy that accompanied the frequent changes of government in the Ukraine. His daughter later recalled an incident during the civil war when they were living in an apartment building in Kiev. Vernadsky, who had begun a biogeochemical laboratory in the Ukrainian Academy, was working quietly at home one day, writing a chapter on the geochemical composition of the biosphere for his manuscript *Living Matter*, when a group of army deserters who had robbed all the apartments on the first three floors of their building burst in. He scolded them for interrupting his work, and they quickly retreated, taking nothing. It may also be that Vernadsky, who was known for his spartan lifestyle, had nothing the bandits cared to steal. It was nonetheless a frightening incident for his family.[30]

During the Red Terror, when the Bolsheviks were in control of Kiev during the spring and summer of 1919, they began taking members of the middle class and elements of the old regime as hostages, executing several hundred of them. Vernadsky thought it prudent to go into hiding. According to his diary, after the murder of the Ukrainian biologist Naumenko, who was one of Vernadsky's research assistants, he began to fear for his own and his family's safety and decided to move from Kiev to a biological research station near Starosele, a village on the Dnieper River near the ancient city of Vyshgorod.[31] Newspaper articles in Kiev confused Vernadsky with another former professor, Bernatsky, an economist who had been minister of finance in the provisional government. These articles questioned why such a former high official in the provisional government was allowed to live freely in Bolshevik-controlled Kiev. While the confusion of names was troubling enough, Vernadsky himself, of course, was a former assistant minister of education in the provisional government as well as one of the founders of the Kadet party, from whose Central Committee he had only recently resigned. Therefore, the danger to him was probably real, whether or not the confusion with Bernatsky was cleared up.

During these months of hiding, Vernadsky's research assistant, a nineteen-year-old student named Theodosius Dobzhansky, walked the twelve miles from Kiev once each week with a knapsack of mail and groceries. The streetcars in Kiev had stopped running, and public transportation out of the city was at a standstill. Dobzhansky, later a world-renowned geneticist at Rockefeller University in New York, had just completed his freshman year at the University of Kiev. Dobzhansky was already a "professor" at a rabfak (workers' university) in Kiev and therefore had a valid identification from the local Bolshevik government, which allowed him safe conduct past the checkpoints on the outskirts of the city. Vernadsky, whom Dobzhansky considered "by far the greatest scientific personality of that time in Kiev," spent his period of hiding productively, determining the chemical composition of various species of plants and animals in the district, and checking their relative weights and frequency of occurrence among the local flora and fauna as part of the research he was doing on the role of "living matter" in the biosphere.[32]

In fact, with his withdrawal from active politics, Vernadsky became more and more deeply involved in his scientific work and related philosophical interests. The civil war was one of his most productive periods of scientific creativity, despite the shortages, hardships, and dangers. Science became for him a way of escaping the horrors of war and revolution and still maintaining an optimistic view of the world. In particular, Vernadsky extended his scientific work in three ways during this period. He began to do concrete experimental studies on the chemical composition of different types of organisms. Two of his assistants, for example, found that bismuth and nickel seemed to exist in every form of living matter. They began to measure the amounts in creatures such as domestic mice. Vernadsky's eventual goal was to create a series of tables for many forms of life, showing the percentages of different elements to be found in each. Here he was following the example of the American geochemist, F. W. Clarke, who had created such tables for various types of minerals.

The second way in which he extended his work was to create a series of concepts for the new science of biogeochemistry. Working in the dense forests around Starosele, he first formulated the concepts of the speed of life (that is, the rates at which different forms of life multiply and spread), the prevalence of life, even in places which seemed most inhospitable to it, the pressure of life, and its adaptability. These were all concepts that became central to his new science and took on more concrete meaning as time went on. Vernadsky in particular wished to give mathematical expression to these concepts.

The third way in which Vernadsky extended his work was by the influence he exerted on several Ukrainian scientists who later became prominent. The best known was N. G. Kholodny, the director of the research station at Starosele and a biologist specializing in the study of iron bacteria, a subject which held a special interest for Vernadsky, since here was a clean example of an organism directly involved in the concentration of an important element. He and Kholodny formed a strong intellectual friendship over the next forty years.[33]

While all this intense intellectual work was going on, civil war was raging in the Ukraine and other parts of the former Empire, and Vernadsky found himself no longer a landowner, losing all his accumulated real estate except for his property in the Crimea, which was still in the hands of the Whites. His estate at Vernadovka was lost early in the revolution, and when he tried to return to his Ukrainian dacha at Shishaki in 1918, where he had done some of his most fruitful thinking in the summer of 1917, he found that the house had been sacked and even the apricot trees had been uprooted and stolen.[34] Interestingly, the loss of so much valuable property found no serious expression in his letters and diary for this period. Perhaps he felt a bit like his daughter-in-law, Nina, when she later wrote: "The revolution brought us many hardships, but at the same time it freed us from the great responsibility of being a landowner in Russia."[35] To be realistic, there was little Vernadsky could have done to save his property, but it was to his credit that he did not become embittered and turn to support of the more right-wing elements in the White movement who took punitive action indiscriminately against peasants in areas where they took power.

Vernadsky was, in fact, far too absorbed by his scientific work and excited by its newness and his potential to make a major contribution to world science. "A year ago in Kiev," he wrote in his diary in 1920,

> I put the question to myself about my situation as a scientist. I clearly confess that I have done less than I could have, that in my intense scientific work there was much dilettantism. I stubbornly did not achieve that which I clearly knew could give brilliant results; I passed by discoveries that were clear to me. Old age arrived and I evaluate my work as the work of a middle scientist with some separate, unfinished thoughts and beginnings, outstanding for their time.

But then he noted that over the past months he had changed this evaluation of himself.

> I clearly began to realize that I was meant to say something new to mankind in my work on living matter . . . and that this is my calling, the obligation placed on me, which I must introduce into life, like a prophet feeling inside himself a voice calling to act. I felt in myself Socrates' demon. Now I confess that these studies can turn out to have the same influence as Darwin's books and in that case, although my essence has not changed, I will be counted in the first ranks of the world's scientists, as if all by accident and conditionally. It is curious that the consciousness that in my work on living matter I have created a new body of knowledge and that it presents another side—another aspect—of evolutionary theory, was given to me clearly only after my illness."[36]

The illness to which he referred was his brush with death when, returning in 1919 from one of his trips to Rostov on the Don, he came down with typhus.

Unable to reach Kiev directly, he headed for the Crimea, which was also under White rule by the government of General Wrangel. His wife and daughter were able to meet him there and nurse him back to health. He decided to remain in the Crimea and to seek British help in emigrating, since his scientific work was now his top priority and he could not make much progress in it, he felt, under the conditions of civil war.

The Crimea had become a refuge for many of Vernadsky's distinguished friends, who taught in a new university in Simferopol. Before long, Vernadsky was teaching a course in Stavropol University, and when the rector of that university died from typhus in 1920, Vernadsky was chosen by the faculty as the new rector.[37] As the Red Army closed in on the Crimea, the British and French began to evacuate much of General Wrangel's government, including George and Nina Vernadsky. But Vladimir, his wife, and daughter decided to remain, since many White officers and scholars were unable to flee and begged Vernadsky to remain as rector of the university. He tried to protect them by issuing many of these White officers phony student identification cards, but the Cheka caught on and began to question Vernadsky.[38] Before long, word came from Moscow that a special hospital railway car was being sent to pick up Vernadsky, his wife, and daughter, as well as the wife and daughter of S. F. Oldenburg, who had been staying in the Crimea. This had been arranged by Oldenburg and the Bolshevik commissar of health, Semashko, a graduate of Moscow University who considered himself a student of Vernadsky. (Vernadsky did not remember him but was grateful to be put under the protection of such an important Bolshevik official.)[39]

Sergei Oldenburg met them when they arrived in Moscow during March 1921, and Vernadsky soon paid a visit to his old laboratory at Moscow University. There he met one of his students, E. E. Flint, who seemed hard at work on an experiment. This pleased Vernadsky, who did not realize that Flint actually was making moonshine (*samogon*) to exchange on the black market for food.[40] Vernadsky soon left for Petrograd to resume his position in the Academy of Sciences. However, he was promptly arrested by the Cheka and thrown in prison. This time it was Oldenburg and N. M. Fedorovsky, his former student, who was now head of all the mining industries in Soviet Russia, who intervened with Lunacharsky and Lenin to have Vernadsky released after three days' incarceration. As Fedorovsky put it in a letter to Lunacharsky on July 16, 1921, "Vernadsky is one of the most noble people of our epoch, one of the last humanists, already of advanced age and weak health." Fedorovsky wrote that Vernadsky's arrest could have heavy consequences for the scientist personally and for the Soviet Russian Republic, since Vernadsky was widely known in world scientific circles, "as a person incapable of political intrigue."[41] Fedorovsky no doubt remembered gratefully Vernadsky's intervention on his behalf before the revolution.

Despite his arrest, Vernadsky was encouraged by what he found in Petrograd. Although grass was growing on Nevsky Prospect and other main boulevards of the former capital and although he had to devote a massive amount of time to the basic necessities of life, the main thing, according to Vernadsky, was that "thought is still alive."[42] The Academy of Sciences had managed to carve out for itself an area of autonomy and to preserve a great deal of independence despite efforts over the previous three years by certain Bolshevik activists to abolish the Academy or at least to bring it under the control of the Bolshevik government. This accomplishment was largely the work of such friends and students of Vernadsky as Oldenburg, Fersman, Karpinsky, Kurnakov, Steklov, and Ivan P. Pavlov (Vernadsky's daughter was a student of Pavlov's, and they had lived across the hall from Pavlov before leaving for the Ukraine in 1917).

All of this seems especially surprising since the Academy's members had greeted the October revolution with hostility. In the first few weeks after the revolution, the General Assembly of the Academy had passed a resolution condemning the Bolshevik seizure of power and supporting the election of a Constituent Assembly to create a parliamentary system. Oldenburg personally expressed his indignation and pessimism at a general meeting of the Academy on December 29, 1917, stating that "it would be cowardly not to look truth straight in the eye and recognize now that the Russian people have failed a great historical test and not stood their ground in a great world conflict: the dark ignorant masses have succumbed to a deceptive temptation and to superficial and criminal promises, and Russia stands on the edge of destruction."[43]

What perhaps is surprising is the great caution and tact with which the central authorities of the Soviet government, such as Lenin and Lunacharsky, but not necessarily local authorities or lower functionaries, treated the scientific community, beginning first with the Academy of Sciences. For example, Oldenburg was not arrested as a former minister of the provisional government, nor apparently even harassed for such published statements as the one quoted above. In fact, he remained the chief administrator of the Academy until his removal by Stalin's government in 1929. He in turn was greatly impressed by Lenin when he made a courtesy call to see the new Soviet leader soon after the Bolshevik takeover.

Although there was a great divide beween this Kadet scholar and a radical intellectual like Lenin, Oldenburg seems to have come partially under Lenin's spell.[44] Lenin's passionate interest in education and science and his skills as a politician particularly impressed Oldenburg. For an establishment intellectual such as Oldenburg, the contrast between Lenin and the last Tsar, Nicholas II— surrounded by people whom the educated classes considered charlatans and ignoramuses—could not have been sharper. Although Soviet sources probably exaggerate the ease and degree with which scientists such as Oldenburg were won over to enthusiastic support of the Soviet government, clearly many prominent members of the Academy were willing to judge the situation for themselves as it developed, rather than actively opposing the new government. Their modus operandi for decades had been one of working within the established governmental system for peaceful change, rather than taking to the streets in order to force change more rapidly. The way their relations with the Soviet government developed was quite consistent with their previous behavior.

Several weeks after the October revolution, at a meeting of the newly established Commission on Education (which later became the Commissariat of Education, or Narkompros), a special commissar for the Academy was appointed, I. V. Egorov.[45] His role appears to have been more to protect the Academy from mob attacks and the chaotic conditions of the time than to interfere in Academy work; and by early 1918 he had faded from the scene. Within a month of the October revolution, three officials of the Academy—the president, Karpinsky; the vice president, the mathematician Steklov; and Oldenburg, the secretary—paid a visit to the new commissar of education, Lunacharsky. No record of their conversation has been found, and it is unclear whether this visit represented more than a courtesy call. Real talks between the Academy and the Soviet government began in January 1918.

The Academy was the first Russian scientific institution that the Soviet government attempted to win over to active cooperation. Although small, with only forty-four full members and 220 employees in 1917 (109 of them qualified specialists), in prestige and authority the Academy represented the acme of Russia scholarly life. Fewer than half its full members in 1917 were in the natural sciences; but in contrast to earlier periods of its history, the natural scientists had begun to dominate its activity by 1917.[46] Karpinsky and Steklov, both natural scientists, became its first freely elected officers after the fall of the Tsarist government in 1917. The Soviet government was to foster this trend toward dominance of the natural scientists in the Academy by its policy of budget appropriations after 1917.

The process of accommodation between the Academy and the Soviet government began as early as January 1918 and continued into the spring. In January a special section of the Commission of Education was created for the "mobilization of scientific forces" to aid the Soviet government. The section was headed by a former Menshevik who joined the Communist party in 1918, L. G. Shapiro. Shapiro promptly met with Sergei Oldenburg and presented him with a request that the Academy aid the Soviet government in matters requiring scientific expertise, particularly in areas related to the economy and social policy. Shapiro's written proposal was that the Academy would create a special commission for this purpose which would draw on the talents of all scientists in the country, not just those associated with the Academy.[47]

According to Steklov's diary entry for January 24, 1918, Oldenburg reported to a general meeting of the Academy that Shapiro had promised the Academy "full preservation of its independence."[48] On January 29, Lunacharsky reported to Narkompros that the Academy was ready to aid Narkompros and the Supreme Council of the National Economy (VSNKh) and had agreed to enter into talks about reform of the Academy.[49] Lunacharsky mentioned nothing about a promise to preserve the independence of the Academy, whereas the published report of the Academy's discussions of this first approach from the government for cooperation makes no mention of the Academy's willingness to consider a reform of its structure. Thus at the very beginning of formal talks between the Soviet government and the Academy of Sciences, two issues surfaced that were to become points of friction between Soviet authorities and members of the Academy. Nonetheless, there were factors driving the two groups together as well: the Academy, always a state-supported institution, needed money to operate and protection from popular forces among the radical intelligentsia and working class that wanted to abolish the Academy altogether, such as the Proletkult movement and the Education Department of the Petrograd regional government (Sevpros).

A general meeting of the Academy discussed Shapiro's proposal on February 16 and formed a commission to prepare a reply. Aleksandr Fersman, the academic secretary of KEPS, wrote this commission a long memorandum in which he objected to giving the Soviet government advice on particular, specialized segments of the economy. The Academy, he felt, should stick to broad scientific questions. He added, "In these grave moments of Russian reality, the task of conserving what we have must take precedence over the idea of new tasks."[50] A few days later, the Academy formulated its reply to Shapiro. The note was couched in vague terms. The Academy, in essence, indicated that it would judge each individual request for help from the government on its own merits and would help in ways where it could and was equipped to do so. The main task of the Academy, the reply emphasized, was to foster scientific creativity, but it would aid the government where possible "for the good of Russia." The Academy's reply was followed a few days later by a request from Oldenburg for an advance to the Academy of 65,000 rubles. On March 5, Lunacharsky wrote the Academy disagreeing that it was not properly equipped to aid the government on economic and technical problems and urging cooperation. On March 9, Oldenburg again wrote Lunacharsky, asking for money for the Academy. This time Narkompros was forthcoming with a general appropriation and also intervened with the local Petrograd authorities to prevent some of the Academy's space from being requisitioned as living quarters for outsiders. Lunacharsky also assured the Academy that its press would not be seized or closed down, a fear that may have been provoked by the actions of the Soviet government in closing opposition newspapers in this period.[51]

By March 24, Fersman had apparently experienced a change of heart and sent a memorandum from KEPS outlining the concrete ways in which it was willing to help the Soviet economy. That same day the president of the Academy, Karpinsky, sent a letter to Lunacharsky indicating that the Academy had not ceased working a single day since the October revolution. At the same time he complained of the popular view in Soviet circles that intellectual workers were somehow privileged and undemocratic. A week later (April 2) Oldenburg again wrote Lunacharsky asking for money, this time specifically to support

KEPS, which he indicated had never had a regular budget under the tsarist government. On April 5, 1918, Lunacharsky publicly recounted this fact in a newspaper article about the Academy, and on April 19, the Soviet government issued it first public decree in the area of science, financing the work of the Academy to study the natural productive resources of Russia.[52]

Although Lenin apparently took no direct role in these early negotiations with the Academy, he was pleased with the initial results. As Lunacharsky remembered Lenin's reaction during the April Sovnarkom meeting, "Vladimir Ilich responded to my report about the Academy of Sciences that it was necessary to support it financially, to motivate it to take those steps which would link its work with our tasks, that it was necessary to find support there among the more progressive scientists."[53] Lenin advised Lunacharsky to publicize widely the fact that the Academy had agreed to cooperate with the Soviet government. "The fact that they wish to help us is good," he reportedly told Lunacharsky: "Tell the whole world that the Academy of Sciences has recognized the government."[54]

In March and April 1918, the Soviet government also began to formulate more general principles as the basis of a science policy. In the March 1918 issue of the journal *National Economy* [*Narodnoe khoziaistvo*], an article appeared under the title "The Mobilization of Science." It was signed with the initials of Lev Shapiro. Since Shapiro was the head of Narkompros's section for the mobilization of scientific forces, it seems likely that this represented an official view from within Narkompros. The article emphasized several points which were to become shibboleths of Soviet science: the need to bring science closer to production, the need for more collective forms of scientific research, and the requirement for centralized state regulation and direction of scientific research. The latter point was especially acute in a country such as Russia, where qualified scientists were few in number and dispersed in their concerns and in their institutional affiliations.[55]

Although no direct reaction by scientists to this article has been found, a close reading of their articles and statements over the next several years indicates that leading spokesmen for the scientific community held some strong reservations about these principles.[56] Many, if not most, favored more application of science to problems of the economy, but those in basic research feared a neglect of more fundamental questions in scientific research. The trend in world science toward more collective forms of research in large institutions and laboratories was undeniable, but prominent spokesmen such as Aleksandr Fersman qualified that emphasis by indicating that much of the work of synthesis in science was still the task of individuals. Their creativity remained a matter of fostering an environment that encouraged individuality, an openness to debate and critical thinking by individuals.[57] Regarding the third point, state direction and planning of science, organized groups of scientists throughout this early period reiterated the need for scientists themselves to determine the direction of science, the need for freedom to organize autonomous groups that would perhaps work closely with the government but would not be subsumed within it.[58]

In April 1918, Lenin outlined some of his preliminary thinking about science and the needs of the economy. His notes were sketchy but constituted his first written attempt at a plan for the reorganization of industry and the economy of Russia. He wanted VSNKh to ask the Academy of Sciences to appoint scientific and technical specialists to a series of commissions which would develop detailed plans for the more rational distribution of industry, closer to sources of raw materials. With the lessons of World War I acutely in mind, particularly in the consequences of the blockage which cut Russia off from Germany and from easy access to the allies, Lenin also outlined the need to develop more self-sufficiency in raw materials and manufactures which previously had been imported. A third emphasis was on a plan for the rapid electrification of Russia.[59] The Academy's and other scientists quickly picked up on these general points and began proposing research programs in new scientific institutes in areas of particular concern to Lenin and the government. Already in April, KEPS began to organize new sections in many of these areas, which later evolved into institutes in the applied sciences: sections to find and study rare elements, new sources of fuel and raw materials like iron, a section on optical technology, where Russia had earlier been mainly dependent on the German optical industry, and so on.[60]

By April 1918, in other words, the Academy of Sciences was eager to show the government its interest in conducting work in such areas of applied science, now that the government had demonstrated its willingness to provide funds and encouragement. Despite this rapprochement between scientists and the Soviet state, evidence of conflict is also abundant in the years 1918–19. Between April and July 1918, the secretary of the Sovnarkom, N. P. Gorbunov, began to conduct talks with large numbers of scientists about the goals of the economy and of Russian science. In June alone, Gorbunov twice communicated to the Academy Lenin's desire for their views about the tasks of Russian science. In response KEPS drafted a memorandum, "On the Tasks of Scientific Construction."

While Lenin and Gorbunov were following a kind of technocratic model in seeking to involve scientists and other specialists in government commissions and as consultants on a variety of practical problems, in April Narkompros and the local Education Department of the Petrograd regional government embarked on a different approach to science, aimed at changing its institutional structure. Narkompros's direction increasingly alarmed not only scientists but Gorbunov and Lenin as well. Narkompros began to work out plans and develop pressure for the reform of scientific institutions and their democratization. The head of the Education Department of Petrograd, M. P. Kristi, even proposed the abolition of the Academy of Sciences, a "completely unnecessary survival of a false class epoch and class society." In March 1918, Narkompros abolished the section for the mobilization of science and replaced it with the Scientific Section (NTO), which had two responsible workers: L. G. Shapiro and its new head, a young astronomer, V. T. Ter-Oganesov, who began to work on a reform plan for Russian science.

Until August 1918, the Scientific Section was the only government bureau that had as its primary responsibility the supervision of Russian scientific institutions and the development of science policy. It was created by Narkompros with the specific charge to make the reform of scientific institutions its first priority.[61] The assistant commissar of education, M. P. Pokrovsky, who took a jaded view of academics and professors from the old regime, was apparently one of the first to suggest, at a meeting of the collegium of Narkompros on April 24, 1918, that this reform take the general direction of creating an association of Russian science, to which the Academy of Sciences would be transferred.[62] According to the diary of the vice president of the Academy, Steklov, by late March and early April 1918, Ter-Oganesov had begun to talk to Academy members about the creation of a "federation of scientific societies."[63] By June 1918, L. G. Shapiro was already discussing the details of such a reform project with Sergei Oldenburg.[64] Ter-Oganesov reported to Narkompros that the association would include "representatives of scientific societies and institutions and would be a body to which the government could turn for the solution of problems in all branches of knowledge."[65] While the advocates of this plan spoke of it as creating a kind of "parliament of scientific opinion," they also intended it to be directed and coordinated by the government and closely linked to the tasks of economic development.

Members of the Academy of Sciences at first treated this plan cautiously. They also began to propose their own reform plans. For example, in June 1918 Fersman sent Narkompros a reform plan which called for the creation of a "Union of Scientific Organizations." The Soviet historian Bastrakova denies that this was a counterplan, but her own analysis of differences between the Narkompros project and the Academy project suggest otherwise. The plans differed in their internal organization and, more significantly, in terms of the new body's relationship to the state. The Narkompros project made it clear that its association would be subordinate to the government, whereas the Academy project foresaw a "Union of Scientific Institutions" receiving financial support from the state but being independent and self-governing. Only science and its workers were considered competent to establish the form and direction of their work and their relationship to the government.

Beyond that, there is a further hint that scientists, particularly in the Academy, were disturbed by the plan and concerned about the fate of the Academy should it be subsumed in such an association. Gorbunov, who had been conducting extensive talks with scientists since April, wrote a letter to the Central Committee of the Communist party in July protesting that the reform plan of Narkompros "would harm an institution of worldwide prestige and hinder active cooperation between scientists and the Soviet government."[66] What provoked this sharp reaction and apparently also hardened the opposition of the Academy was a new variant of the Narkompros plan produced in the summer of 1918 which clearly threatened the existence of the Academy and its traditions. In January 1919, the president of the Academy, Karpinsky, wrote directly to Narkompros warning of the dangers contained in its reform plan. In this same period, Lenin apparently had several conversations with Lunacharsky, telling him that reform of the Academy must await "a quieter time" and warning him against breaking any "valuable china" in the Academy. Members of the Academy, Lenin indicated, had shown a cooperative attitude toward the Soviet government, and Narkompros's plans threatened disruption of that relationship.

What probably happened is that prominent members of the Academy, increasingy alarmed by what they were hearing from officials in Narkompros, attempted to find patrons for their interests by appealing to prominent Bolsheviks outside Lunacharsky's bailiwick. Gorbunov was readily available and was close to Lenin. He was also a young chemical engineer (he had received his degree in 1917) impressed by the authority of scientific graybeards, with whom he established good working relations in subsequent years.[67] Although it is unclear whether the concern of Academy members is what motivated Gorbunov's letter to the party's Central Committee, this may have been what happened, judging from another instance in which scientists sought to reach Lenin's ear.

On August 15, 1919, Sergei Oldenburg wrote to the well-known physicist P. P. Lazarev, who at that time was conducting field studies in the Ukraine. He pleaded with Lazarev to use his influence with the prominent Bolshevik engineer, L. B. Krasin, to gain Lenin's help. Krasin headed the state's Commission for Supply of the Red Army and used this important position to aid Lazarev's research with supplies and other assistance. Krasin, a long-time associate of Lenin, was in frequent contact with the Soviet leaders during the civil war, and Oldenburg hoped that he might intervene in the Academy's behalf. As Oldenburg wrote Lazarev,

> A black cloud from Moscow, they say, is hanging over the Academy: Artemev and Ter-Oganesov [officials of Narkompros] have some kind of plan for the complete liquidation of the Academy simply by decree. No one and nothing, of course, can abolish science while even one person is still alive, but it is easy to disorganize. Talk with Krasin, ask him to speak with Lenin, who is an intelligent man and understands that the liquidation of the Academy of Sciences would bring shame to any government.[68]

Whether Krasin spoke directly with Lenin about this issue is unclear from the record, but in 1919 Lenin intervened to stop the Narkompros plan. He called in Lunacharsky several times to express his concern, telling him, according to Lunacharsky's account published in 1925, that "it is not necesssary to let the Academy be devoured by a few Communist fanatics." Lunacharsky says that he defended the need to adapt the Academy to existing government institutions and prevent it from remaining "a state within a state" but agreed with Lenin that the Narkompros plan was not appropriate or timely.[69]

Lenin did not fully trust former "left Bolsheviks" such as Lunacharsky and Pokrovsky because of their long-time association with the radical cultural policies of the Proletkult movement, and this probably heightened his concern about their aims toward the Academy.[70] The leaders of the Academy for their part proposed a different reform plan to Narkompros in 1919. In July 1919, Narkompros rejected the Academy's plan as too mild and "not conforming with the spirit of the times."[71] Nonetheless, much of the Academy's plan was eventually instituted. In general by 1920, the Academy of Sciences had become a more open and democratic institution than the old Imperial Academy, and it remained relatively autonomous from the Soviet government, although completely dependent upon it financially.

During the civil war period more than forty new scientific research institutes were created, more than half of the seventy which existed by 1922. Besides those under the jurisdiction of the NTO, a number were organized in various commissariats, including the Commissariats of Agriculture and Health. The former palaces of the Romanovs and the mansions of the nobility and upper bourgeoisie were often turned over to such research complexes, and, although severe shortages of virtually everything needed for research greatly hampered work, many scientists were clearly impressed by the rapidity with which so many of their proposals were accepted and instituted by various branches of the government. Applied research became centered in the NTO and in commissariats which supervised branches of the economy outside of industry. Institutes in areas of more fundamental research generally remained under the Academy of Sciences, or were created by Narkompros and placed under its jurisdiction. For example, the X-Ray Institute and the Optical Institute, in which much fundamental research on the structure of matter was conducted under the auspices of A. F. Ioffe, were created in this period and placed under Narkompros.[72]

Thus, the period did not see the creation of a monolithic, highly centralized system of scientific research but one that remained flexible and largely decentralized. While opportunities grew for scientists to display initiative and realize a number of projects they had only dreamed about before 1917, the leaders of the scientific community remained, with good reason, concerned about long-range Soviet plans for science, and they were determined to keep as much corporate autonomy and room for maneuver as possible.

Prominent natural scientists such as Aleksandr Fersman, while hailing the creation of many new institutes, also believed that they might contain some dangers. Fersman, for example, warned against over-specialization, fearing that institutes composed exclusively of specialists in a single field could lose the broader view of a problem and he therefore recommended that such scientific organizations be composed of a variety of scientists who could communicate their differing knowledge and points of view to each other. Fersman also feared that too much emphasis on collective research might work to the disadvantage of individual creativity. He warned:

> I see the future of scientific creativity in the harmonious combination of these two paths [individual and collective research], and it will be destructive if either one of them triumphs by itself. . . . For different natures, for different minds, there cannot be and must not be identical forms of creativity, and woe to that organization which would wish to impose such. Let collective creativity develop and let the individual mind work freely.[73]

Most projects put forward by scientists during this period contained demands for autonomy, demands which were always rejected in principle by Soviet authorities but often tolerated in practice.[74] Scientific workers were also the last major professional group to be organized into a Soviet-controlled union, something which did not occur until 1923. Through the civil war, significant elements of the scientific community were members of an independent union that lobbied for their interests and attempted to improve living and working conditions. Prominent individual scientists also played a vital role in preserving the independence of scientific institutions and improving living and working conditions for scientists. For example, some time toward the end of 1919, I. P. Pavlov, the Nobel prize-winning physiologist, wrote the Soviet government asking permission to leave Russia so that he could continue his scientific work abroad. Lenin and his secretary V. D. Bonch-Bruevich, felt that this would be a serious blow to Soviet prestige abroad, where hostile propaganda emphasized the uncivilized nature of the Bolshevik regime and the decline of science and education under its rule. When Lenin saw Pavlov's letter he blamed the Petrograd regional government for not looking after the needs of such eminent scientists. Bonch-Bruevich was instructed to write Pavlov assuring him that the Soviet government would provide everything he needed to continue his work. Pavlov's reply to Bonch-Bruevich's letter was a detailed description of the difficult conditions in which Petrograd scientists were forced to live and work. Lenin instructed his secretary to reply that the government would take prompt measures to improve the situation of scientists. Bonch-Bruevich again wrote Pavlov requesting that he postpone his emigration, promising that the government would move rapidly to help scientists.[75]

Nonetheless the government was slow to act. A protocol approved by the General Assembly of the Academy warned of the grave situation in which Russian scientists found themselves. This protocol was sent to Lenin and the Sovnarkom early in December 1920 and resulted in a meeting between Lenin and prominent members of the scientific community in January 1921. There he assured them that measures would be taken to improve their living and working conditions. But on May 17 the Academy again complained that little had been done. If immediate measures were not taken, the Academy threatened, Russian scientists and their families should be allowed to go abroad, "where their health and lives will be preserved for scientific work." Thus the Academy accompanied its demands with a veiled threat that if nothing were done, demands for a large-scale emigration of Russian scientists would take place. They probably knew from Pavlov's experience that top Soviet leaders would want to avoid such an embarrassment, especially as they sought diplomatic recognition and respectability in Western Europe and America.

Although Vernadsky arrived late on the scene in Petrograd, after the end of the civil war, he very quickly set to work organizing new institutes and commissions in his area of interest. It was not possible to do much substantive experimental scientific research in the conditions of that time, but it was possible to engage in institution building. In 1921 and 1922, before leaving for Paris to work with Marie Curie, Vernadsky organized the Radium Institute, based on his former laboratory in the Academy, and sent memoranda proposing the creation of a commission on the history of knowledge, the later nucleus for the Institute for the History of Science and Technology, and a polar commission to study the 40 percent of Russia's land mass that was covered with permafrost. As director of the Academy's Mineralogical Museum, he also created a section for the study of meteorites headed by his associate, L. A. Kulik, who promptly organized an expedition to Siberia to study the mysterious Tungus explosion of 1908.[76] Although Vernadsky found it difficult to conduct his own scientific research in a city that had received no new Western scientific literature in several years and was deprived of many of the elementary necessities of life, he was nonetheless impressed by the accomplishments of his friends and colleagues who had remained in Petrograd during the civil war. In March 1923, then living in Paris, Vernadsky wrote to his longtime friend I. I. Petrunkevich, now a political exile:

> Let me touch on the state of science in Russia. It seems to me that here [in the West] they do not recognize the huge cultural task that has been accomplished, accomplished in the face of sufferings, humiliations, destruction. Scientific work in Russia has not perished, but on the contrary is developing. . . . scientific work in Russia has been preserved and lives a *vast* life thanks to a conscious act of will. It has been necessary and will be necessary to fight for it every day, every step. . . . I tell people [here] how this has been accomplished and how much has perished. People have suffered much—and in the Ukraine, perhaps in connection with the chauvinistic policy of the Ukrainian Bolsheviks, the universities in particular [have suffered], but their scientific life has been preserved. The Russian Academy of Sciences is the single institution in which *nothing* has been touched. It remains as before, with full internal freedom. Of course, in a police state this freedom is relative and it is necessary to defend it continually. Much new has been created in Moscow and Petersburg, *de facto*

much, although by comparison with the plans of 1915–17, little. And curiously enough, much has been created in the provinces.[77]

Vernadsky, no friend of Bolshevism, gave credit for what had been accomplished primarily to the scientific intelligentsia itself, and although it is clear that much of the initiative and effort to preserve and develop science had come from scientists, it is also clear from the record that such accomplishments could not have taken place without the active support and cooperation of the central Bolshevik authorities. This was particularly true of Lenin and the Sovnarkom, Lunacharsky and Narkompros, Gorbunov and VSNKh, and the heads of many other central commissariats, such as the commissar of health, Semashko, who needed science and who protected the natural scientific community against their more zealous colleagues in the Communist party and local Soviet. In fact, one of the major conflicts that surfaced during the revolution and civil war and remained an important issue in later years was the tension in Soviet culture between egalitarianism and a technocratic approach. On one side there was a popular movement to do away with privilege, among whose representatives scientists were frequently counted by the masses and by local party and Soviet activists. On the other was the more technocratic approach favored by central Bolshevik authorities such as Lenin, who believed in the necessity of expertise and who established institutional means for consulting experts, providing in return protection and certain privileges. The community of natural scientists proved to be effective in exploiting to their own advantage such conflicts within the Soviet government and Bolshevik party, in order to prevent the centralization of science and protect the relative autonomy of many of their institutions.

When one compares the aims and accomplishments of the Soviet government toward science and the aims and accomplishments of scientists themselves, the scientists appear to have had the advantage by 1921. As early as 1918, the Soviet government aimed at creating a centralized system of scientific research, planned and controlled by the government, not by scientists. Scientists preferred a highly diversified and decentralized system controlled by their own autonomous organizations. Although they did not achieve such control, by 1921 scientists enjoyed considerable influence in a diversified and decentralized system of governmental research organizations, in which there was little planning or central coordination, a system that was much more the product of institutional conflicts (for example, between Narkompros and VSNKh) than it was of Marxist-Leninist ideology. In terms of the historiographic debate over the relative importance of ideology and pragmatism during the period of war communism, this study provides ammunition for those who argue the case for the pragmatism of Soviet decisions. If ideology had an influence, it was primarily the ideology of the leaders of the scientific community, not Bolshevik ideology. Although Vernadsky played no direct part in preserving the values of the scientific intelligentsia in Soviet Russia during the civil war period, his friends and students, such as Fersman, were among the leaders in this movement, and when Vernadsky returned to Petrograd in 1921, he was pleased at what had been accomplished. When he left for Paris in 1922, it was with a promise to M. P. Pokrovsky (deputy commissar of education, in charge of scientific institutions) that he would return and play an active role in the development of Soviet science.[78]

CHAPTER
5

THE VERNADSKY SCHOOL AND SOVIET SCIENCE, 1922–1945

When Vernadsky and his family arrived in Paris, they found the way already paved for them by their émigré friends, in particular A. V. Golshtein and V. K. Agafonova, who had arranged an apartment for them and obtained for Vernadsky at least one year's support from the Sorbonne to teach a course on geochemistry. He also had a promise of some support from Marie Curie and her institute.[1] A four-year struggle for the soul of Vladimir Ivanovich ensued. His wife and émigré friends did not want Vernadsky to return to Soviet Russia and accused Sergei Oldenburg and other friends who had chosen to remain in Russia and work for the Bolsheviks of putting undue moral pressure on him to return. Golshtein was strongly opposed to such "conciliators" and was herself an "irreconcilable," similar to many among the French middle class who had invested heavily in the old regime in the hopes of building up a strong ally for France and who had then lost everything when the Bolsheviks came to power. For Golshtein, the Bolsheviks were little better than the "Antichrist," and when she saw that Vernadsky was likely to return to the USSR eventually, she let him know that she considered him to be on the opposite side of the barricades.[2] Vernadsky himself had a different view of the "conciliators." He later compared such friends as Oldenburg and Fersman to "certain Chinese mandarins who had saved Chinese culture under Chingiz-Khan."[3]

At the same time, the Academy of Sciences through Oldenburg and Fersman accused Vernadsky of intending to stay in the West and refused to renew the authorization for his trip beyond one year. Worse than that, they canceled his salary, although they left him his title of Academician. Vernadsky protested that he needed more time to complete the work he had begun in Paris and was furious at the lack of trust in him displayed by the Academy. He understood how precarious was the position of the Academy and how his lengthened stay in the West might complicate matters for the Academy. He understood the struggle that was going on in Russian culture, but he had to put his work first, given his advanced age. He considered it his right to stay wherever he wished as long as he wished. If that meant losing his position with the Russian Academy then he wanted it known that this was the Academy's doing and not his. He still

160

believed it would be easier to organize his long-term work with colleagues in Soviet Russia than with those abroad, and he had no desire to become an émigré.[4]

Nonetheless, evidence from this same period in the Bakhmeteff Archive at Columbia University indicates that Vernadsky did in fact make a considerable effort to remain in the West and to obtain permanent funding there for his new scientific interests.[5] However, he was unable to obtain such funding, and in 1925 the Soviet authorities tried to encourage his return by awarding him a newly created chair in the Academy. Along with the chair, the Soviets promised Vernadsky the freedom and the funding to pursue his own line of research.[6] The Soviet government had a keen appreciation for both Vernadsky's enormous scientific prestige and his moral authority. As one of fewer than fifty members of the old Imperial Academy of Sciences, his return and cooperation with the Soviet government would add legitimacy to the new regime and help to attract the old intelligentsia whose experience and expertise were desperately needed for Soviet reconstruction efforts.

On Vernadsky's part, whereas his age (sixty-three) and new scientific interests worked against him in the West, in the Soviet Union he could command enormous respect and greater material resources, especially now that the Soviet economy had begun to recover. Beyond that, he hoped to attract bright, enthusiastic young people who would carry on his efforts once he had departed.[7] Vernadsky was impressed by the enthusiasm for science in the USSR and by the enormous creative intellectual and moral potential of the Russian people. By 1925, it should be noted, many members of the old intelligentsia had become impressed by the moderation of the New Economic Policy launched in 1921 and the material recovery and somewhat greater freedom it had stimulated in Soviet society. A considerable movement to cooperate with the Soviet government had grown among some émigrés as well as former opponents who remained in the USSR. Vernadsky personified at this time an important tendency within an element of the intelligentsia. This was a belief in the pursuit of science as a major (if not the major) goal for human intellectual activity and as the best hope for the future of the human species. It was this belief that provided a bridge for many Russian intellectuals between the old society and the new. Some of those who, like Vernadsky, disliked Bolshevism were able to persuade themselves that Bolshevism was ephemeral and that they could pour the new wine of science, secular culture, and economic development into the old wineskin of a Marxist movement they believed had lost its dynamism.

For example, in a 1925 letter to Fedor Rodichev, Vernadsky expressed the view that the "Utopia of Communism," which was perishing ideologically, was not dangerous. What he saw rising in its place was a growing economy, a growing nationalism, an increasing interest in science and moral ideals, all of which he considered favorable developments. The best, most talented young people, he felt, were not attracted to the ideology of communism (or socialism) but to these other movements.[8] Particularly to those intellectuals who, like Vernadsky, did not believe in the efficacy of violent revolution and who also rejected a religious view of human life, science as an activity had great appeal. In this sense, there was a convergence of interests with Soviet authorities. In the areas of science and technology, the Soviet authorities in the mid-1920s were trying to win over the so-called old specialists, extending a policy begun by Lenin and others during the civil war.

For this combination of reasons, therefore, and despite his ambivalence, Vernadsky returned to the USSR in the spring of 1926. The Soviet government made it almost as difficult for him to return as it had been for him to leave in the first place, probably more from the sheer inefficiency of the system than from any malice aforethought. His letters from this period are filled with complaints that his visas had been delayed and that his salary from the Academy had not been received. For a man who before the revolution had been comfortably well off, Vernadsky was reduced to begging from various friends and institutions enough money to buy warm winter clothing for the journey to Leningrad. In the meantime he accepted whatever lecture dates and stipends he could obtain in France and Czechoslovakia, where his son and daughter were living.[9] Nonetheless, by 1926, he had given up the search for a permanent place in the West and was ready to return to an environment where he felt he could concentrate on his work and in which there would be more continuity with the past, more assurance of support in the future.

Notwithstanding the excitement of living abroad and keeping up with the latest European developments, such as the work of his friend V. M. Goldschmidt (the famous geochemist at Heidelberg), the American geochemists at the Carnegie Institution in Washington, and physicists associated with Niels Bohr and his Copenhagen school,[10] Vernadsky had used his time in the West very well. Besides conducting important research on radioactive minerals for the Curie Institute (funded by the Rosenthal Foundation), he published two important books in French, La Géochimie, his synthesis of the new science of geochemistry based on his lectures at the Sorbonne, and La Biosphère, his pathbreaking book defining the concept of the biosphere and outlining what was known and what needed to be known in this area. His lectures at the Sorbonne were attended by two young French scholars, the Jesuit scientist Teilhard de Chardin and the French philosopher Edouard Le Roy. They borrowed from Vernadsky his usage of the term biosphere and he in turn borrowed from Le Roy his idea of the noosphere, that is, the idea that the biosphere was being transformed into a geological zone controlled by human reason, a kind of scientific utopia that became for Vernadsky a key element in his thought during the last fifteen years or so of his life.

Despite these successes in popularizing his ideas in the West (he had always felt that the fact he published his work mostly in Russian had served as a barrier to the acceptance of his ideas outside of Russia and was now determined to overcome that by publishing his major work in French as well), Vernadsky and his wife, low on money, returned to Leningrad in March 1926. Vernadsky brought in as his assistant for the new lab the talented young scientist Aleksandr P. Vinogradov, who had begun to study with him in 1918. (Vinogradov later became vice president of the Academy of Sciences and a foremost Soviet oceanographer as well as the chief Soviet space geologist in the 1960s and early 1970s.[11])

Back in the USSR, Vernadsky did not become a silent, passive follower of the contemporary Soviet line. To the contrary, he took a leading part in the struggle to preserve the autonomy and prevent a Communist takeover of the Academy of Sciences after 1927, opposing the election of Marxist philosophers such as A. M. Deborin to the Academy. Vernadsky defended his stand in a long memorandum in which he asserted that Marxism was an "outworn philosophy" that held back the development of the sciences.[12] While this memorandum has never been published, Vernadsky did publish controversial views on Marxism and the relationship between science and philosophy several times during the 1930s in the publications of the Soviet Academy of Sciences and the Communist party. The standard Soviet interpretation of Vernadsky in the 1930s was that he was a brilliant scientist but a poor politician and philosopher.

A. M. Deborin, one of the Marxist philosophers whose admission to the Academy of Sciences Vernadsky had earlier tried to block, set the tone for the polemic in an article from 1932: "The whole world view of V. I. Vernadsky actually is deeply hostile to materialism and to our contemporary life, our construction of socialism."[13] Deborin accused him of being an eclectic in philosophy who never criticized religion or idealist philosophy in his works. In fact, he suspected Vernadsky of being sympathetic to religious philosophy.

Vernadsky replied in a subsequent issue: "I consider myself obliged not to leave Academician Deborin's article without a reply."[14] Vernadsky accused Deborin of attempting to hinder his scientific work within the borders of the USSR by making fantastic accusations and using methods of criticism that weakened "the scientific work of our country." What Deborin mistook for eclecticism, Vernadsky added, was actually philosophical skepticism. He was trying in his work on biogeochemistry to free science, and biology in particular, from nineteenth-century views, which he considered distorted by philosophical conceptions not based on experiment and observation of nature. Such experimentation Vernadsky considered the only basis for scientific knowledge: "I am a philosophical skeptic. This means that I consider that not one philosophical system (including our official philosophy) has achieved that eternal applicability which science (and only in several specific areas) has achieved."[15]

Vernadsky protested against those critics who tried to pin labels on him, quoting from his earlier writings in order to place him in some philosophical cubbyhole: "My world view has not been static from the 1880s to 1916 and later."[16] Vernadsky considered that his view point since 1916 had been one of philosophical skepticism:

> As a result of his investigations, Academician Deborin comes to the conclusion that I am a mystic and the founder of a new religious-philosophical system;

others have defined me as a vitalist, a neovitalist, a fideist, idealist, mechanist, mystic. I must precisely and decisively protest against all these definitions. I must protest not because I considered them offensive to me but because they are false in relation to me and spoken lightly by people who are talking about something they know nothing about.[17]

Vernadsky went on to indicate that in a country whose growth and development depended on scientific thought and work, "scientists must be saved from the tutorship of representatives of philosophy."[18] On the contrary, philosophers had more to learn from science, since Vernadsky considered that "all old philosophical constructs do not embrace the new, quickly growing scientific description of reality."[19]

Deborin's reply was to assert the right of philosophers to criticize the views of scientists. Vernadsky, he wrote, "is sufficiently condescending to the poor philosophers to allow them to drink the ambrosia and nectar from his scientific table but only let them restrain themselves from criticism."[20] Deborin quoted from some of Vernadsky's earlier writings, republished in 1922, to point up his sympathy for mysticism and religious thinkers.[21] Earlier, Vernadsky had written that philosophy, religion, and science expressed different aspects of the human spirit and that these three spheres were not necessarily in contradiction. Vernadsky's fault, in Deborin's view, was clearly his liberalism, his tolerance of mysticism or religion and of an idealistic philosophy not necessarily harmful to science. In fact, Vernadsky pointed to the religious quest as the basis of the work of such great scientists as Newton and Maxwell. There is nothing in the record to indicate that Vernadsky ever changed his views in this regard, despite his philosophical skepticism and belief that only scientific knowledge, based on empirical generalizations derived from observation and experiment, can have universal binding force for all human beings. Religion and philosophy, Vernadsky believed, are subjective and necessarily vary from individual to individual.

Vernadsky's real offense, then, was to deny that dialectical materialism was the only valid philosophy, binding on all persons in a socialist society and useful to scientists in their professional search for truth. "I can only have the audacity to think," Deborin wrote, "that if Academician Vernadsky did not treat so carelessly and condescendingly 'our philosophy' but would occupy himself with a serious study of the classics of Marxism-Leninism, then perhaps he would disavow his archaic views, hindering his scientific work; and having learned the method of dialectical materialism, would find in it a better tool of scientific research."[22]

Deborin, on the other hand, denied ever trying to hinder Vernadsky's actual scientific work: "In my critique I did not touch on his concrete scientific work in the sphere of biogeochemistry where Vernadsky has substantial achievements. Considering his 'empirical' scientific work very valuable and important, I, of course, could not and cannot have any of the 'diabolical intentions' which Vernadsky attributes to me."[23]

Vernadsky continued to find opportunities in his published writings of the 1930s to criticize contemporary philosophy, including dialectical materialism. When he did so, he generally provoked a sharp reply. For example, in an Academy of Sciences journal in 1937, he published a scientific treatise on the limits of the biosphere, in which he criticized official philosophy in the USSR as old-fashioned and scientifically outmoded.[24] The Soviet philosopher A. A. Maksimov was quick to reply:

> In his article on the boundaries of the biosphere, Academician Vernadsky has also touched on the contemporary status of philosophy in general and in the USSR in particular, a question having nothing to do with the biosphere. He is free to do so, but his method and answers stand in sharp contradiction to the methods of scientific work and to the methods of scientific work of Vernadsky himself.[25]

Maksimov went on to contrast the favorable conditions for scientific work existing in the USSR with those in the capitalist countries. Not once, Maksimov continued, does Vernadsky say anything about idealism and its negative effect on the development of science in those countries:

> In capitalist countries scientists struggle with religious, mystical, idealistic, and other anti-scientific tendencies. In the USSR at the present time, the path has been swept clean for science. The social roots of religion and idealistic philosophy have been cleared. So all the necessary conditions have been created for a full flowering of science, the beginnings of which are already seen. This is what dialectical materialism has given science.[26]

Hundreds of scientists in the USSR, Maksimov maintained, already had become supporters of dialectical materialism, and the Academy of Sciences was the first such scientific institution in the world to include people in its ranks studying ways to apply dialectical materialism to science. Yet Maksimov admitted that the work of Marxist philosophers still lagged behind the successes of contemporary science, and this "doubtless has a negative influence both on the development of science and the development of philosophy."[27]

Vernadsky was not one to let philosophers such as Maksimov have the last word. It has now become clear that in the final ten years of his life he was concerned with leaving as a legacy for future generations a scientific world view that would be an alternative to Stalinist philosophical dogmatism. Vernadsky's career shows a consistent willingness to criticize the established authorities, whether Tsarist or Stalinist, while working within the system for peaceful evolutionary change.[28] While he cooperated with the Soviet industrialization drive and was obviously impressed by the enthusiasm for science and education, he was also just as obviously in conflict with Stalinism regarding the constraints it placed on the free development of science. Not only was he critical of the purges, which directly affected personnel in his laboratory; he also criticized the state of Soviet science and its lack of greater freedom of thought and creativity.[29] Some might say that his criticism tended to stop short of any real sacrifice of his scientific interests. For example, as his son notes, he ceased his opposition to the reorganization of the Academy of Sciences in February 1930, although he refused to retract his opinions on the election of Marxist philosophers to that institution. Vernadsky was allowed to remain head of his laboratory and continue his productive research in geochemistry, for which he received high state honors before his death, including the Stalin prize. As he wrote at the age of eighty (in 1943): "It is scientific work and research, independent scholarly thought, and the individual's creative search for truth which has occupied and continues to occupy the foremost place in my life."[30]

In the years following his return to the USSR until his death in 1945, Vernadsky refrained from organized political activity. Nonetheless, he became known in the late 1920s and the 1930s not only as a critic of Marxism but also as a protector of non-Marxists. Among other things, he provided refuge in his laboratory for the children of persons persecuted for their social origins and their politicial and philosophical views, including the daughter of his friend Prince Dmitrii Shakhovskoi and the son of the scientist and Russian Orthodox priest, Pavel Florensky, who were arrested in the late 1930s. Anna Shakhovskaia became Vernadsky's secretary and preserver of his memory and Konstantin P. Florensky, who came as an orphan to Vernadsky's lab at the age of eighteen, became one of the USSR's leading space geologists. Vernadsky helped a number of such people in a variety of ways, including financial aid, petitions to the authorities, and letters of recommendation.[31] That Vernadsky was a man of independent views who was quite open in expressing himself, even at the height of the purges, seems well established. Given what we know of his strong personality, this fact is not as surprising perhaps as is the official Soviet reaction to his dissent. His rejection of Marxism-Leninism as a useful guide for Soviet science was, of course, criticized in the press, as we have seen, but none of this criticism led to his arrest or even a mass campaign against him.

Vernadsky and his close scientific associates were not untouched by Stalinism, but because of a particular combination of circumstances they did escape the worse fate of other scientists, such as the geneticist Nikolai I. Vavilov and his school. For one thing, Vernadsky was a highly respected mandarin, a graybeard whose prestige among the scientific intelligentsia was enormous. He was a link with the past of the Academy of Sciences, one of only forty-five full members of the old Imperial Academy at the time of the 1917 revolutions and one of the Academy's most respected members. He had, after all, been trained by such luminaries of Russian science as Mendeleev and Dokuchaev and was close to the inner circle of the Academy; he had been a friend of Pavlov and of Karpinsky, the prominent geologist and president of the Academy until his death in 1935. Vernadsky was respected both for his scientific contributions and for his active struggle against the abuses of tsarism, although he had been a leftist liberal before the revolution and not a Marxist. His philosophical views could be and were dismissed as survivals of the past, the eccentricities of an old man who was nonetheless very useful in training new scientific personnel and advancing the accomplishments of Soviet science. The Soviets valued his authority highly. That he had chosen to return to the Soviet Union voluntarily and to work peacefully for the Soviets were useful facts in persuading other non-Marxist scientists to work for the regime.

A second important reason for the survival of Vernadsky and his school—and perhaps an even more important one in the atmosphere of growing nationalism during the 1930s—was Vernadsky's well-known Russian patriotism. He stressed his patriotism in private communications to such persons as Viacheslav P. Volgin, the chief administrator of the Soviet Academy in the early 1930s, and Viacheslav M. Molotov, the Soviet premier after 1930 and one of Stalin's closest associates.[32]

Third, Vernadsky was a scientist of international renown whose arrest or persecution would not have helped the Soviet reputation with so-called progressive world opinion, particularly in the Western scientific community. Vernadsky went abroad every summer until the mid-1930s and had a wide circle of scientific acquaintances in the West, where he published frequently in professional journals, gave lectures, and attended conferences.[33]

A fourth important reason for his school's survival was the fact that Vernadsky was not known as an intriguer or plotter but as a man of integrity who was always open in his disagreements. The manner in which he criticized, the limits he observed in his criticism, and the nature of his opponents may have helped him survive and may have protected his school. Most of the published attacks against him came from Marxist philosophers such as Deborin who were in trouble with Stalin, especially after 1932. It was not uncommon for Stalinists to worry more about Marxists with whom they disagreed and whom they distrusted than they did about non-Marxists who worked loyally for the regime, did not intrigue, and were no real threat to Stalin's position. Vernadsky's published dissent in the 1930s was directed at the validity of dialectical materialism as a helpful tool of scientists and did not go beyond that to a full-blown critique of the political or social system. His remarks usually appeared in limited-circulation publications of the Soviet Academy, and he made no attempt to publish his dissent abroad. His most severe criticisms of Soviet science were not published at all in these years but were found in memoranda and private letters directed through proper channels to Soviet officials. Vernadsky was careful not to embarrass these officials publicly and thereby earn their enmity, a major departure from the way in which he openly criticized tsarist officials in the public press before 1917.

Last, and perhaps most important for ensuring survival, Vernadsky had been known for some years as one of the Academy's strongest advocates of combining theoretical studies with applied science. This emphasis on applied science, particularly science for the needs of the economy and national defense, clearly appealed to one of the strongest Stalinist biases about the role of science and scientists in this period. Vernadsky's advocacy of applied science was not simply rhetorical. His scientific research and that of his school were valuable to the regime in a variety of ways, which can be documented from Soviet sources. Vernadsky and his school made many practical contributions to Soviet economic and military efforts in the 1930s and during World War II. These practical contributions, while important, formed only a small part of the scientific activity of the Vernadsky school, but they did lend protective coloration to the rest of the school's scientific enterprise, which was often more theoretical in nature.

Furthermore, Vernadsky's theoretical work often had practical implications. With his work on the paragenesis of minerals—the influence that different minerals have on each other's formation—he formulated a theory of isomorphic series that indicated which minerals tend to be found together, a theory that was not an infallible guide to geological prospecting but did provide clues for Soviet scientists who were undertaking a more systematic scientific search for natural resources in the 1930s. Since minerals, and ways to find those most valuable for Soviet industry and agriculture, were seen as keys to Soviet power, it is not surprising that the Soviet leadership would take the work of Vernadsky and school very seriously.

Among his practical contributions in the 1930s, Vernadsky helped found the scientific study of the permafrost that covers more than 40 percent of Soviet territory, and he headed the academy's first Commission on Permafrost.[34] In the 1930s Vernadsky also retained a strong interest in radioactivity, studies of which he had helped to institutionalize by founding the Radium Institute in 1922.[35] He and his students founded the field of radiogeology in the USSR, studying the effect radioactivity has on geological processes. During the 1930s, they began to map the radioactivity of the Soviet Union's surface and tried to determine the age of geological strata by using radioactive methods.[36] Vernadsky was particularly concerned to locate Soviet deposits of radium and other radioactive elements, since their importance for medicine and industry was already recognized. In 1930 Vernadsky published an important article on the concentration of radium by living plants, and in 1932 he chaired and helped organize the first All-Union Conference on the Phenomenon of Radioactivity.[37]

In the same year, Vernadsky and his student Khlopin, at that time the assistant director of the Radium Institute, began to build the first cyclotron in the Soviet Union. They encountered a number of problems in completing this effort, including insufficient material support and technical difficulties that greatly slowed the effort and delayed the actual operation of the cyclotron. These problems contributed to the large gap between the theoretical and experimental side of research on radioactivity in the Soviet Union. Nonetheless, some of the Soviet Union's leading atomic physicists were trained on the cyclotron, including Igor V. Kurchatov, the man who eventually led the project building the Soviet Union's first atomic weapons after World War II.[38]

In 1933 Vernadsky learned that a United States scientist, Gilbert Lewis, had created heavy water. He understood the significance of this development for nuclear energy research and for medical and other biological studies using tracer elements and proposed that a study of heavy water be undertaken in the Biogeochemical Laboratory (BIOGEL). On November 6, 1933, the Soviet chemist Nikolai S. Kurnakov chaired a conference in the Academy on the question of organizing Soviet research on heavy water. On May 22, 1934, the academy created a commission on heavy water with Vernadsky as its head. Vernadsky's BIOGEL became the site for the construction of an apparatus for making heavy water, and the apparatus was in operation by 1935.[39]

In the late 1930s Vernadsky's laboratory developed particularly close ties with the government ministries responsible for health and agriculture and fulfilled a number of research projects for them. From 1935 to 1938, for instance, BIOGEL looked for ways to extract useful rare elements from asphalt, oil, and coal. It also performed research on processes in which organisms participate, such as the formation of nitrates useful for fertilizers and the role of isotopes in various organisms. In formulating its five-year plan for 1938 to 1942, BIOGEL sought to intensify its research on the biogeochemical role of rare elements and their possible uses in agriculture and medicine.[40]

Scientists in Vernadsky's laboratory studied chemical deficiencies or excesses in the environment and the effects of imbalances on the health of local inhabitants. In 1935, a group of medical workers turned to Vernadsky for his help in diagnosing the causes of an illness affecting people in the Lake Baikal area. In this physical condition, called the Uroskii endemia, the bones of children and young persons became deformed, making them invalids. Vernadsky's laboratory sent eleven researchers in several expeditions to this area to study the chemical composition of the waters, soil, and mineral deposits. They concluded that such endemic illnesses were restricted to certain biogeochemical areas or provinces, as the Vernadsky school called them, and that they resulted from the environmental lack, or oversupply, of certain chemical elements, such as iodine, strontium, barium, and calcium.

These conclusions were first presented by Vernadsky and Vinogradov in a paper read at a meeting of the Moscow Therapeutic Society on June 5, 1936, entitled "Biochemical Provinces and Illnesses." In a series of works over the next fifteen years Vinogradov developed the idea of biogeochemical provinces and worked on studying and mapping these provinces and on the prevention of endemic diseases related to the chemical environment of a particular area. That such diseases almost disappeared from the USSR in subsequent years was in no small part due to BIOGEL's expeditions from 1930 to 1938.[41]

From 1939 to 1941, Vernadsky was very interested and active in research on applications of isotopes. World science in the 1920s and 1930s, had focused a good deal of attention on this topic, but the Soviet Union lacked a center for concentrated research on isotopes. Vernadsky thought such a center essential, and on June 2, 1939, he proposed to the Division of Chemical Sciences of the Academy that the Commission on Heavy Water be transformed into a Commission on Isotopes.[42] They agreed and asked Vernadsky to publish popular articles in the Soviet press on isotopes and their importance. Vernadsky did so but could not resist the opportunity to criticize Soviet conditions, within the limits of what was considered acceptable criticism. In a 1941 article he wrote that Soviet mining practices were extremely wasteful. He noted that in working ores of potash, uranium, rubidium, and cesium, miners were losing pure isotopes of lead, strontium, and barium. "In our state socialist structure this is impermissible," he wrote, adding in a political vein: "This is understandable under capitalism, with its chaotic production founded on competition, but in a correctly planned economy and with a scientific reckoning of the future, a socialist country must not follow a harmful policy, the final loss of a great deal of matter valuable for human use—now and in the future."[43]

Following in Vernadsky's footsteps, in 1937 Vinogradov published a major part of his fundamental work on the chemical composition of sea organisms and established his reputation as one of the Soviet Union's leading oceanographers.[44] By the start of World War II, BIOGEL was recognized as a highly productive and creative part of the Academy of Sciences and had increased from only ten or so scientific researchers in the late 1920s to more than thirty qualified scientists by the beginning of World War II.[45] By 1940 in the Soviet Union as a whole several hundred scientists, according to Soviet sources, were working on problems first formulated by Vernadsky. Vernadsky's reputation for dissent had not prevented the growth of his own laboratory or of the number of his followers, although general social and political conditions of the 1930s did hinder their efforts in other ways.

One last example of Vernadsky's scientific work in this period needs to be mentioned, since it was of great importance for the Soviet Union's transformation into an atomic superpower after World War II. In 1940 Vernadsky took an active part in the creation of the Uranium Commission, and during World War II he agitated for a crash program in atomic energy. In 1938 the German scientists Otto Hahn, Lise Meitner, and Fritz Strassmann split the atom, opening the possibility of a chain reaction. Vernadsky learned of this event only after some delay and was alarmed by the Soviet censorship of information from abroad.[46] At the first All-Union Congress on Isotopes, held in April 1940, he pushed for the exploration of uranium supplies in the USSR not only as an urgent need of Soviet science but for national security reasons as well. In June 1940, he received a letter from his son, who now lived and taught in the United States. George Vernadsky enclosed a *New York Times* clipping of May 5, 1940, that discussed the possibilities of using energy released by chain reactions and summarized research in this direction in the United States and elsewhere.[47] This news greatly aroused the older Vernadsky, particularly since it was unclear whether the Soviet Union had an adequate program in this area or even enough suitable uranium ore for research and development. Expeditions were needed, he felt. He immediately began to contact leading geologists, geochemists, and physicists in the Soviet Union, particularly his students Khlopin and Vinogradov, the radiologist Shcherbakov, the geologist and government official Andrei D. Arkhangelsky, and the head of the Institute of Physical-Technical Problems in Leningrad, Abram F. Ioffe, who was considered the dean of Soviet physicists in these years. [48]

In June 1940 Vernadsky and Khlopin sent a memorandum to the Geological-Geographical Section of the Soviet Academy of Sciences, urging the immediate dispatch of expeditions to various parts of the USSR in search of uranium deposits. Vernadsky, Khlopin, and Fersman were asked by the Academy to work out a list of measures, and by June and July 1940 *Pravda* and *Izvestiia* were already reporting on a series of expeditions dispatched to Central Asia in search of uranium. [49] Vernadsky and his associates thought that the most likely place to find suitable uranium ore was in the Fergana Valley of Central Asia. On July 1, 1940, Vernadsky wrote to Otto Iu. Shmidt, a scientist who was close to Stalin and was vice-president of the Soviet Academy of Sciences:

> According to information just received by me almost by accident and in incomplete form due unfortunately to the artificial obstacles placed in the way of our reading the foreign press, it has become clear that in the U.S.A. and Germany energetic and organized work is proceeding in this direction [the development of atomic energy] despite the world military situation. Our country must under no circumstances stand aside from these efforts but must provide the resources and opportunity for broadly organized and rapid work in this area, which has the utmost significance. [50]

A day later, Vernadsky wrote his close friend from student days, the literary scholar Ivan M. Grevs:

> I consider that discoveries made recently, chiefly in America and Germany but to a lesser extent here, open before mankind an enormous future—the use of the energy within the atom, which in intensity and capacity leave far behind both steam and electricity. But this, of course, is a matter for the future. Now it is necessary to discover the means—very great—for finding supplies, mining uranium, and extracting from it the isotope 235. [51]

Ten days later, Vernadsky, Khlopin, and Fersman sent a memorandum to Nikolai A. Bulganin, the vice-premier of the Soviet government who was also president of its Council on the Chemical and Metallurgical Industries:

> We consider that the time has already arrived for the government to consider the importance of solving the problem of the technical use of energy within the atom and to take a series of measures which would assure the USSR the possibility of solving this most important problem. [52]

They proposed building a plant to process uranium, giving the Academy of Sciences the means to speed up its work on building a high-energy cyclotron in the Physical-Technical Institute and creating an adequate reserve of processed uranium.

The same day they addressed a similar proposal to the president of the Soviet Academy of Sciences, with an additional request to convene a second All-Union Conference on Radioactivity at the Radium Institute in Leningrad during the winter of 1940–41. This letter also requested that Fersman be placed in charge of a brigade of scientists who would visit the major uranium sites in the USSR during the fall of 1940 and then hold a small conference in Tashkent on how best to develop these sites.[53] On July 16 the president of the Academy responded favorably to Vernadsky's proposals, but Vernadsky felt that many scientists and government officials did not feel a sense of urgency.[54] Vernadsky wrote in his diary on July 17, 1940: "The large majority of people do not understand the historical significance of this moment. I wonder if I am wrong or not?"[55]

The president and presidium of the Soviet Academy met on July 30, 1940, and adopted Vernadsky's proposal that a uranium commission be created within the academy. Vernadsky was asked to be its chair, but because of his age he requested that his student Khlopin, head of the Radium Institute, be chair instead. Vernadsky and Ioffe were elected vice-chairmen. There were fourteen members of the commission, including most of the leading figures of Soviet physics: I. V. Kurchatov, Sergei I. Vavilov, Petr L. Kapitsa, Petr P. Lazarev, Aleksandr N. Frumkin, Leonid I. Mandelshtam, Iulii B. Khariton, Vernadsky's students Vinogradov, Fersman, and Khlopin, and the geologists Arkhangelsky and Shcherbakov.[56] By August Vernadsky had learned that secrecy had been thrown around atomic energy research in both the United States and Germany. He saw this development as proof of the need for a crash program in the USSR.[57]

The military implications of this effort are clear, but Vernadsky tended to emphasize the long-range importance of the peaceful uses of atomic energy. There is some evidence of conflict among members of the Uranium Commission over how to proceed with research in this area. According to David Holloway's account, Kurchatov and Khariton were dissatisfied with the commission's broad research program and in August 1940 addressed the academy's presidium with a plan for focusing research on achieving a chain reaction as soon as possible, "proposing that an experimental pile be built, using natural uranium and a good moderator."[58] Holloway argues that this plan was opposed by Khlopin and the senior members of the Uranium Commission and that young Turks such as Kurchatov and Khariton, lacking the power to force through their plan, were stymied: "The issue resolved itself into a classic disagreement between a group of younger scientists (Kurchatov and Khariton were in their mid-thirties, while Flerov, Petrzak, Rusinov, and Zel'dovich were in their twenties) who had little reputation or institutional power and a group of senior scientists (including Khlopin, Vernadsky, and Ioffe) who had larger reputations and powerful positions within the academy."[59] Holloway's published evidence for such a generational conflict, however, rests on one passage in a Soviet historical novel in which Khlopin, at the Moscow Nuclear Conference of November 1940, supposedly criticized Kurchatov for being carried away by the immediate practicality of developing a chain reaction and using atomic energy. No corroborating evidence has yet surfaced for Holloway's suggestion—in an otherwise superb account—of a simplistic dichotomy between young Turks and older, more established scientists. Holloway's conclusion does not seem convincing in its argument that the early history of Soviet atomic energy research "offers a classic example of the problems of innovation in scientific research, with the younger scientists seeking to make their reputations in a field that held the promise of major discoveries, while the older, well-established scientists were unwilling to take risks which might harm their position."[60]

In fact, much of the new evidence about Vernadsky, unearthed since Holloway's article was published, suggests that Vernadsky, like Kurchatov and the younger scientists, was frustrated by the inertia of the Academy and the government in recognizing the importance of research in atomic energy. That there was resistance and opposition to large-scale investment in such research in the early 1940s seems certain, but the sources of this resistance apparently did not break down into a simple opposition between generations of scientists. Vernadsky and possibly other members of his school were among those who wished to push ahead rapidly but on a broad front. By the time of the German invasion in 1941, however, Vernadsky was elderly—intellectually alert but in feeble health and unable to carry through as actively as before in his efforts to focus the government's attention on the atomic energy problem. We have several indications that he was disturbed at the lack of Soviet progress in the area during the early years of World War II, and there is evidence of a conflict between Vernadsky and Ioffe, one of the most powerful physicists on the Uranium Commission.[61] In his correspondence with Fersman in November 1942, after other elderly Academicians had been evacuated from Moscow to a resort in Kazakhstan, Vernadsky wrote: "How goes the business of the Uranium Commission? It seems to me urgent that it be active. Khlopin has written that Ioffe went to the government with some kind of memorandum on this subject, completely stifling the efforts of the Academy of Sciences."[62] In a paper written in 1984, Holloway suggested that Ioffe was especially cautious about the prospects for atomic energy research because he had suffered a major embarrassment in the early 1930s when he had promised a new insulating material that would greatly improve electric power transmission and was then unable to deliver on his promise.[63] Ioffe apparently thought that the development of atomic energy would take many decades and probably would not be practical until the twenty-first century.[64] We do not yet know the positions other major figures on the Uranium Commission held on the question of atomic energy—for example, Sergei Vavilov and Kapitsa, who were major scientific powers in their own right. Until further archival sources become available, therefore, it is difficult to speculate on the issues and disagreements among Soviet scientists on this question, other than to suggest that Vernadsky shared the impatience of younger scientists with the lack of a Soviet crash program. Nonetheless, Vernadsky and his associates may have differed with the younger scientists on the exact nature of such a program and on whether it should explore a broad range of problems or focus narrowly on creating an atomic pile and achieving a chain reaction and uranium bomb as soon as possible.

We know that the secret Soviet atomic weapons program began to gear up in early 1943, although the precise chain of events leading up to that program is much less certain. At the end of 1942 Kurchatov, one of the physicists who had worked in the Radium Institute under Khlopin and then in Ioffe's Leningrad Institute, was placed in charge of the Soviet effort to create a uranium bomb and, in February or March 1943, began work on the bomb project in Moscow.[65] Vernadsky apparently was not privy to this secret project.[66]

At any rate, in 1941 and 1942 the work of the Uranium Commission definitely lessened, in part because of the war but also, in Vernadsky's view, because certain scientists, especially Ioffe, underestimated the importance of atomic research. Vernadsky was upset by delay and expressed his concern in a number of letters to Vinogradov, Fersman, and others. In a letter of March 15, 1943, to Vladimir L. Komarov, president of the Academy of Sciences, Vernadsky wrote that "unfortunately Ioffe does not understand or pretends not to understand that for the use of atomic energy first it is necessary to find uranium ores in sufficient quantity. I think that in one summer campaign this problem may be solved. As far as I know, Fersman and Khlopin are of the same opinion."[67] In a note a few days earlier, Vernadsky had commented that little progress in exploration for uranium ores had been made since the mid-1930s: "In this instance, the state of our knowledge is the same as it was in 1935. Our huge bureaucratic apparatus has proved powerless."[68] In April 1943, Komarov replied favorably to Vernadsky's entreaties, indicating that his memo had been sent on to the council of ministers, apparently with some effect, since exploration for uranium resumed later in 1943 with at least one of Vernadsky's close associates, the geologist Shcherbakov actively involved.[69] Thanks in part to the efforts of Vernadsky and his school, an active program of uranium prospecting had been undertaken before the end of World War II and became a vital contribution to the Soviet nuclear weapons effort.[70]

What can we say specifically about the effects that social and political conditions under Stalin had on Vernadsky and his school in Russian science? On the plus side, Vernadsky's laboratory, BIOGEL, more than tripled in size during the 1930s and the Vernadsky school grew considerably in influence, as Soviet science itself developed both institutionally and in numbers during this decade. Vernadsky was able to found a new subfield, biogeochemistry, which he had begun to develop after 1916 but which had few researchers before 1928. It became a flourishing endeavor only in the 1930s, despite the difficult times. Vernadsky also was generally pleased with the progress of Russian geochemistry, the academic and research field he founded in Tsarist Russia. It became well established under his student Fersman in the 1930s. Studies of radioactive minerals also advanced in these years under the aegis of Khlopin, Shcherbakov, and others close to Vernadsky and Fersman. The resources made available by the government for scientific research in such areas were far more generous than they had been in earlier decades.[71]

Vernadsky was less pleased with other developments of the 1930s, particularly with the general situation in mineralogy, the quality of many young scientists, the censorship of information and difficulties with foreign travel after 1935; the inertia of Soviet bureaucracy and its overcentralization, the lack of effective scientific planning, and, not least, the arrests of talented young scientists, the waste of their talents, and the disruption caused by these arrests.[72]

In mineralogy Vernadsky was very critical both of the quality of specialist education and of the people in charge of research. "A basic reform is necessary and a significant change of personnel," he wrote to Fersman in 1941.[73] He criticized the lack of modern equipment and adequately trained personnel, particularly the narrowness of their specializations, their frequent inability to think independently, and their parochialism. They "do not even know what is happening in world science. . . . It seems to me, when I look at this work [in mineralogy] that it is a morass. People simply do not know what contemporary science is like. . . . They spend huge amounts of money—and the results are insignificant."[74] Vernadsky urged that Soviet mineralogists, as well as geochemists, be trained thoroughly in chemistry, a lack he felt acutely throughout the 1930s. Some of the graduate students assigned to him by the academy he dismissed as no better than "mechanics and windbags."[75]

After 1935, it became increasingly difficult for scientists such as Vernadsky to travel abroad, while the censorship of scientific information from other nations increased. Whatever his own patriotic feelings for Russia, Vernadsky believed strongly in the internationalism of science and did not think that Soviet science could flourish without close contacts with the West. He was particularly annoyed that even popular scientific journals such as *Nature* and *Scientific American* often arrived months late and with large segments cut out by the censor's razor blade.[76]

Vernadsky apparently was able to publish most of his purely scientific works during this period, although not without some harassment and delays from the authorities. He wrote Fersman on March 7, 1933, that the first volume of his massive study on the geochemical significance of water had finally appeared after a delay of a year and a half. The second volume had already been typeset for months and he had still not received the page proofs: "People jump about, play at dialectical materialism (a substitute for philosophical thought), but they do not work. It is a great shame that Russian scholars have not learned how to behave properly. . . . Scientific work right now is close to a catastrophe."[77]

Several of Vernadsky's longer works, which contained philosophical discussions, were held up by the censors or not published at all until after his death. A lengthy article entitled "Goethe as Scientist," which Vernadsky wrote in the mid-1930s for an anthology of the German thinker's works, was held up by the arrest of the anthology's editor in 1936 and was not published until 1946, a year after Vernadsky's death.[78] A volume entitled *Biogeochemical Essays* was ready for publication in 1935, but Soviet philosophers objected to several passages. Vernadsky refused to change them, and the book did not appear until 1940, with the offensive passages intact but with an editorial note disagreeing with them added to the volume. Vernadsky was forced to petition Glavlit, the Soviet censorship bureau, a number of times, both for himself and for others. He wrote in disgust to a friend after one of these encounters that "the censors are ignorant young people."[79]

Vernadsky believed the situation in genetics to be especially damaging for Soviet science. He commented on this belief a number of times in his private correspondence, as in a letter of 1944: "Unfortunately, all of our centers of scientific work in genetics have been destroyed. In the selection of scientific personnel to the Academy of Sciences, I see the pernicious effects of this mistake by the government."[80]

Vernadsky's own scientific school escaped the wholesale purges that destroyed the ranks of Soviet genetics, but his associates and students did not remain untouched by arrests in the 1930s. One of his closest friends and associates, the earth scientist Boris L. Lichkov, was arrested in the early 1930s and spent most of the decade working as a prisoner-specialist for the NKVD on various construction projects until his release late in the 1930s.[81] Lichkov had been Vernadsky's assistant as director of the Commission for the Study of Natural Productive Resources (KEPS) from 1927 to 1930, until Vernadsky was ousted from that position in 1930 and replaced by Academician Ivan M. Gubkin. Gubkin, a party member, was a geologist whom the Vernadsky group considered to be something of a careerist and to be less talented than many non-party scientists.[82] Vernadsky kept up a lively correspondence with Lichkov throughout the years the latter was a prisoner and helped Lichkov obtain the academic credentials he needed for a university teaching job after his release. Much of this correspondence has now been published in the Soviet Union in two volumes, although highly censored and without mention that Lichkov was under arrest during most of the 1930s.[83]

In November 1936 at least three young scientists from among some thirty scientists working in the BIOGEL were arrested, although we do not know a great deal about the reasons for, or circumstances surrounding, their arrests. Two of them, Kirsanov and Lebedev, disappeared forever. The third, Simorin, a medical doctor, was released in 1957. Vernadsky had tried to obtain the release of all three. After their arrests, Vernadsky had written to the NKVD demanding to know their fates and how to correspond with them. When he received no answer, he wrote to the president of the Soviet Union, Mikhail I. Kalinin, whom he knew personally, and eventually obtained their prison addresses. Vernadsky had wanted to write Stalin to protest such arrests, but a delegation from the Academy of Sciences visited him and begged him not to. There is evidence that Vernadsky did write more than one petition to the Supreme

Soviet and other Soviet agencies, protesting the innocence of the three scientists and the disruptive effects of arrests on scientific work.

One of the most serious effects of such arrests on the Vernadsky school was that Vernadsky himself became much more formal with, and distant from, younger scientists who shared his interests and whom he feared endangering by his outspokenness. As the purges of the 1930s intensified, he quit inviting such persons to his home and put them at a distance, warning, in the memory of one survivor, "I cannot change my views and you may suffer from this."[84] Such circumstances scarcely created the kind of supportive atmosphere that encourages a close interchange of views. It is a credit to those of the Vernadsky school that they accomplished as much as they did in these years.

Despite some praise of Stalin and the war effort in his diary during World War II, Vernadsky at the same time sharply criticized the state of Soviet science in memoranda to the academy and in letters to friends and fellow scientists. For example, in 1943 he wrote to Vladimir Komarov, president of the Academy, that, whereas the Soviet military had modern equipment, the Academy did not. "For this it is necessary first of all to create an Institute for Preparing Necessary Equipment and Instruments. This has been discussed for ten years, yet nothing has been done."[85] In a letter to Fersman about the same time, Vernadsky wrote that he was in favor of the planning of science, "but among us there is no such realistic planning."[86]

In general, Vernadsky felt during this period that Soviet science was in need of a major overhaul. During the war, he called for its reconstruction, for autonomous organizations of scientists, for the widening of scientific knowledge and research, and for the assurance of full freedom of development for all scientific schools of thought (a position particularly pertinent to Lysenkoism and similar schools that claimed exclusive access to scientific truth and sought to repress their opponents).[87] "Above all," he wrote to the mathematician N. N. Luzin in June of 1943, "we must increase our scientific capabilities, the strength of our science. In this respect, we are exremely backward and bear a burden for the talent of our people. I consider that we must change this side of our life in a fundamental way."[88] In a letter to the president of the Soviet Academy in 1943, Vernadsky urged particularly close ties with the United States after the war: "It seems to me that we need to enter into much closer contact with American scientists. I consider that at the present time American scientific organization and scientific thought stand at the forefront. We must turn for help to America in reconstructing after the depredations of Hitler's vandals."[89]

Vernadsky also thought that the merging of smaller institutes into centralized institutions that tended to dominate studies in a particular field was a harmful tendency in Soviet science. He felt that this trend was especially pernicious when persons he considered scientifically incompetent were put in charge of such institutes.[90] A greater decentralization of science, a freer flow of information, the end of the state monopoly on the import of foreign publications, and the right for contending schools to exist and debate without fear of reprisals remained throughout this period an ideal for Vernadsky—an ideal that Soviet science was far from fulfilling despite the "scientism" of the regime and the far greater material resources invested in science under Stalin in comparison with earlier periods. Among the most severe limitations Stalinism imposed on Soviet scientists was the prohibition of debate, either public or private, about the social and political conditions of Soviet science. There is no evidence that any of the serious criticisms Vernadsky voiced in private letters and memoranda to Soviet officials during this period were seriously considered or acted upon, although by voicing them he helped to keep alive a critical tradition in Soviet science.

Although to some extent the Vernadsky school was able to function creatively, advancing scientific knowledge and its applications in these years—at times against great odds—among the harms imposed by Stalinism was the atmosphere of intimidation that led to the suppression after Vernadsky's death of several major works he had written during the 1930s and World War II. One of these, *The Chemical Structure of the Earth's Biosphere and Its Surroundings,* which Vernadsky considered the summation of his life's work, gathered dust in the archives until 1965, when it was published in censored form.[91] Another, *The Thoughts of a Naturalist,* was published in the late 1970s in two volumes, again in abridged form.[92] This latter work, even in its censored form, is clearly an attempt to show—through the history and philosophy of science—the contrast between Stalinist and other authoritarian ways of thinking and a scientific world view in which all knowledge is regarded as partial, tentative, and incomplete. In other words, it attempts a useful antidote to the certainties and claims to exclusive truth of Marxism-Leninism.

Vernadsky wanted to demonstrate in these writings how scientific knowledge grows and changes through an atmosphere of intellectual freedom with the right of opposing groups to argue, debate and seek adherents without the fear of intervention by church or state, political party, or state security police. Even more than his scientific discoveries, this attitude was the legacy he wished to leave to his country and his scentific heirs. It is a legacy worthy of his long and original career.

CHAPTER

6

THE LEGACY OF VERNADSKY'S SCIENTIFIC AND PHILOSOPHICAL THOUGHT

Vladimir Vernadsky's scientific and philosophical legacy spans fifty years of Russia's most volatile social and political change. Considered one of the major progenitors of the science of ecology and geochemistry in the Soviet Union, he pioneered studies of biogeochemical cycles in nature, wrote the first scientific treatise on the biosphere, and saw the biosphere being transformed during this century into a region of life dominated by human activity and its products. His life work has been continued by the Russian schools of geochemistry and biogeochemistry, which he founded, and his pupils and scientific heirs dominate Soviet science in this area up to the present.[1] Vernadsky's scholarly legacy, however, contains much more than the body of his scientific research. Never one to focus exclusively on a specific topic or area of research, Vernadsky pursued broad scientific and philosophical questions. Ultimately, to Vernadsky, thinking scientifically meant putting complex natural phenomena into the form of laws, which would manifest themselves over time. Well read in a variety of current scientific and philosophical debates, including evolution, the theory of relativity, atomic theory, and radiation research, Vernadsky proved particularly adept at synthesizing material from different scientific disciplines into comprehensive hypotheses concerning natural phenomena. Vernadsky's coworkers and students often claimed that the issues and scientific agenda posed in his articles comprised Vernadsky's greatest contribution to Soviet science.

The later 1950s witnessed a popular revival of Vernadsky's scientific and philosophical thought. A group of scholars at the Vernadsky Institute decided informally to study and prepare the previously unpublished works of Vernadsky. Renewed interest in Vernadsky has been due in part to an increased awareness of environmental issues in the Soviet Union but, more importantly, due to an attempt to revitalize and attract the Soviet intelligentsia by providing an alternative materialist world view to traditional Marxism-Leninism. Vernadsky's writings from the twenties and thirties appear particularly appropriate for this task. His professed reliance on observable and verifiable data, his optimism

179

concerning scientific and technological progress, and his belief in the inseparable nature of man and his environment remain fundamentally materialist, while not necessarily constrained by the dogma of Marxist historical materialism,

Vernadsky placed vital importance on the primacy of scientific thought as a catalyst to the philosophy and world view of any particular era. An avid aficionado of the history of science and the history of human knowledge in general, Vernadsky saw the twentieth century as an era of scientific creativity comparable to the scientific revolutions of the seventeenth and eighteenth centuries. In Vernadsky's words, the current "explosion" in scientific thought was historically unique in that new theories of atomic structure affected all branches of science simultaneously and thereby changed the basic picture of the cosmos.[2] According to Vernadsky, scientific thought did not exist apart from individuals, and, throughout time, talented individuals created ideas which changed the basic world view of others. In doing so, they effectively changed the biosphere itself, both in a physical sense and in the way man perceives his relationship to the cosmos and the world around him. Vernadsky's emphasis on the value and contributions of individual minds who assimilate and create new ideas from the debates in the scientific community at large attests to his belief in the need for an open community of scientific thought and debate.

In his last years, Vernadsky was concerned with leaving, as a legacy for future generations, a scientific world view that would be an alternative to Stalinist philosophical dogmatism. In fact, he saw this work as a critical step in the emergence of the noosphere, that era when human reason and technology would become vital instruments for the formation and changes occurring in the universe.

Beginning in 1935, at the age of seventy-two, Vernadsky wrote a series of essays on science and the relationship of scientific knowlege to philosophy.[3] In these essays, most of them unpublished until the last decade, he made clear his belief in the superiority of scientific knowledge, arrived at through empirical scientific methods, over any form of philosophy, valuable though philosophy might be in formulating questions and hypotheses and providing a rigorous logic. This was just the opposite of the Stalinist insistence on the superiority of Marxist philosophy, in particular the Stalinist interpretation of dialectical materialism over science. While in these years Vernadsky read widely in philosophy and was willing to concede to Marx, Engels, and Lenin (he pointedly omitted Stalin) their value as philosophers and politicians, he was unwilling to concede to Marxism-Leninism a monopoly of philosophical method or speculation.[4] Vernadsky considered all philosophy valuable as evidence of human consciousness, but scientific knowledge he viewed as superior to any form of philosophy in its value to human beings.[5] In this sense, he pitted himself against Marxist-Leninist dogma and the strong drive on the part of some Soviet philosophers to creat a Marxist-Leninist philosophy of science to which all practicing Soviet scientists should adhere.[6]

Vernadsky steadfastly refused to label himself a dialectical materialist, instead describing himself as a "cosmic realist," a term used earlier by Mendeleev. In fact, Vernadsky accepted the reality of the physical universe outside himself and believed the most valid knowledge to be that obtained from its scientific study.[7] Vernadsky believed mankind's greatest hope lay in the development of science and the spread of scientific thinking and knowledge among the masses, a goal he had pursued throughout his career. Another goal he pursued throughout his lifetime was to free scientific research from external restraints. "One of the preconditions of the contemporary revolution," he wrote during World War II, "is freedom of scientific thought and scientific research, freeing it so far as possible from the pressure of religious, philosophical, and political structures and creating within the social and political system conditions favorable to free scientific thought."[8]

Vernadsky was in many ways the archetype of the twentieth-century scientific organizer, at the hub of a network of scientific institutions and collegial relationships, and a team leader of unusual talent, with a much praised ability to inspire his associates.[9] Vernadsky was also very much a part of the nineteenth- and twentieth-century scientific tradition, which relied on observable and empirical fact yet placed undying faith in the omnipotence of human reason and man's ability to discover the immutable laws of nature. Vernadsky was one of the first scientists to emphasize the basic unity of earth, humans, and the cosmos through the exchange of matter. To Vernadsky, human beings were first and foremost inhabitants of the planet, one component in a cycle of physical and chemical interactions and transmutations, and as such they possessed an obligation to think and act for the good of the planet as well as for their own personal comfort and well being.[10]

Contemporary Soviet environmentalists have sought to use Vernadsky's enormous scientific authority to justify the promotion of the environmentalist movement in the USSR, particularly with the increase in such awareness since the early 1960s.[11] At the same time, Soviet official optimism about the continuation of material progress through industrialization and an increase in human rational control over nature can also find strong support in Vernadsky's work. Yet both of these instances reveal more about the traditional Soviet penchant to legitimize current positions in a public debate by quoting selectively from some highly respected earlier authority than it does about Vernadsky's essential ideas or career.

In the period since 1960, with the growth of an environmentalist mentality and movement, Soviet interest in Vernadsky's ideas and career has undergone a significant revival, including a substantial increase in articles on his career as well as the publication of several major manuscripts of his, written in the 1930s and during World War II, which had been gathering dust in the archives for decades.[12] Beyond that, several new editions of his major works relevant to ecology and first published in the 1920s and 1930s (for example, *The Biosphere* and *Geochemistry*) have been republished.[13] The connection between this resurgence of interest in Vernadsky and a Soviet concern for the environment has been made explicit a number of times in Soviet publications. An example is the following statement in a 1967 edition of *The Biosphere* by the noted Soviet scientist, Dr. A. I. Perelman:

> Vernadsky saw what a huge geological force humanity had become, how quickly it was transforming the planet, how it was changing in a basic way the migration of chemical elements. He emphasized that man is artificially creating processes which never before existed in the biosphere and are alien to it. . . . He issued a call for the study of these phenomena from the view point of geochemistry, to analyze them, to study the long-range consequences of economic activity.
>
> All the ideas of this scientist have now become more relevant than at the time of their publication. In our time geochemical analysis is necessary for the solution of the problems of the conservation of nature, the struggle against water pollution, soil erosion, atmospheric pollution, the disposal of radioactive wastes and so forth. The ideas of the founder of geochemistry will continue to illuminate the path of research of these important problems for years ahead.[14]

We need to go beyond the contemporary uses made of Vernadsky's work in the USSR, to probe beneath the legend of Vernadsky and analyze the genesis of his thinking about the biosphere and its relationship to his life and the social environment of Russia between 1917 and 1945.

The origins of Vernadsky's concept of the biosphere and his research in biogeochemistry are to be found in the period just preceding and following the Russian revolution. But the general scientific attitudes that shaped his approach to this subject were formed earlier, particularly in his work as a student of V. V. Dokuchaev, one of the founders of modern soil science.[15] Dokuchaev emphasized both the historical and holistic approaches to natural science. Dokuchaev studied soils both in terms of their historical evolution and in terms of the interconnectedness of their constituent parts and the relationship of soils to the rest of the natural environment in which they are found. Vernadsky applied this approach first to crystallography and the study of minerals, helping to move from a largely descriptive science in which classification into typologies and the mapping of deposits predominated, to an historical approach which emphasized the processes of mineralogical formation in the earth's crust and the paragenesis of minerals.[16] Vernadsky later wrote, in 1935:

> While reading mineralogy at the university (in St. Petersburg) I began on a path at that time unaccustomed, particularly in connection with my work and contacts during my student years and immediately afterward (1883–97) with the great Russian scientist V. V. Dokuchaev. He first turned my attention to the dynamic side of mineralogy, the study of minerals through time. . . . This defined the whole course of my teaching and study of mineralogy and was reflected in my thought and the scientific work of my students and colleagues.[17]

It was Dokuchaev who, in 1898–99, posed the problem to which Vernadsky later devoted much of his life:

> Looking more attentively at the great discoveries of human knowledge, discoveries, one might say, which have revolutionized our view of nature from top to bottom, especially after the work of Lavoisier, Lyell, Darwin, Helmholtz, and others, it is impossible not to notice one very real and important shortcoming. . . . They have studied chiefly separate bodies—minerals, mining deposits, plants and animals, and individual phenomena—fire (vulcanism), water, earth, air, in which, I repeat, science has achieved astonishing results, but not their interrelationships, not that genetic, eternal and always orderly link between inert and living nature, between plant, animal, and mineral kingdoms on the one hand and man, his daily life and even spiritual world, on the other. But it is expressly these interrelationships, these lawful interactions that comprise the essence of a knowledge of nature, the core of a true natural philosophy—the best and highest achievement of scientific knowledge.[18]

Not one existing science, Dokuchaev continued, can fully embrace these ties. Their study must become the object of a new science—the science of the future.

Vernadsky, in his famous lectures of 1902 on the development of a scientific world view, echoed his teacher's words on the need for a unified view of nature.[19] He noted that the idea of the interconnectedness of all nature is found in ancient religions and philosophies (a fact familiar to him both from his own extensive reading in religion and philosophy and from discussions with his close friend from university days, Sergei Oldenburg, who became a noted Orientalist and specialist on Hindu civilization in the Academy of Sciences). But this idea could remain in science, he argued, only if it survived the test of scientific method and was given a scientific formulation in laws that could be verified. Vernadsky's approach to the question of the origins of life on earth provides a pertinent example of his scientific world view. In 1931, Vernadsky acknowledged that the appearance of life on earth was a subject usually dealt with in religion, philosophy, and art. Vernadsky, however, proposed approaching the issue through the narrower study of the biosphere—more specifically, defining those conditions necessary to sustain life. He redefined the issue of the origins of life as the problem of the origin of the biosphere, thus bringing the debate closer to factual reality and enabling scientists to draw on the empirical evidence presented by geology and geochemistry.[20]

Vernadsky's holistic approach to science goes a long way toward explaining the seemingly diverse nature of his interests and research. He advocated studying major scientific questions in an interdisciplinary manner and repeatedly sought to redefine disciplines in physico-chemical terms. This approach led to the creation of new branches of scientific research such as geochemistry, biogeochemistry, hydrogeochemistry, and hydrogeothermia (the study of underground water, its structural and chemical changes, and its interaction with minerals which together make up the thermal dynamics of the earth's crust), and to continued emphasis on the need to synthesize knowledge in an era of growing specialization.[21] The success of Vernadsky and his students in the practical application of the new sciences to the location of natural resources, agronomy, animal husbandry, and medicine added legitimacy to interdisciplinary research in the eyes of a Soviet government deeply concerned with agricultural and industrial growth.

The other major intellectual influence on Vernadsky which prepared his mind for later work in biogeochemistry was the new chemistry and physics of the late nineteenth and early twentieth centuries. He worked with Mendeleev and, as a post-graduate in Germany and France, became conversant with advanced theoretical work on chemical elements, the interrelationship of such elements, and the nature of chemical bonding. He was fascinated by the implications of the new physics for the earth sciences. As early as 1908, he recognized the significance of radioactivity in explaining many phenomena on earth. He was attracted to this field by one of its pioneers, the Anglo-Irish professor at Trinity College, Dublin, John Joly, whom he met at a conference sponsored by the British Association for the Advancement of Science in 1908.[22]

On the importance of the new physics, Vernadsky was later to write:

> . . . the atom . . . the smallest indivisible body of ancient philosophy and of recent science—this atom has turned for us into a complex system of tiny bodies—a system in which we can study and measure the space, size, motion, and number of bodies composing it, and some of its other qualities. The atom has turned out to be connected in the closest form with energy. . . .[23]

Vernadsky came to the conclusion that the new concept of the atom reflected not only its mass, energy, and spatial extent but also its length of existence and migrations through space, that is, its natural history.[24] The developing conception about the atom prepared Vernadsky to look at life in a new way, from the standpoint of the migration of atoms and their particles within living matter and between living and inert matter.

Between 1916 and 1923, a change occurred in Vernadsky's thinking that was to have revolutionary implications for the way he viewed nature, living matter within nature, and, finally, the place of humanity in the natural environment.[25] Russia's technical and economic unpreparedness for World War I and Vernadsky's scientific work for the war effort stimulated his thought in this direction. The collapse of the old regime and the reshaping of social relations, accompanied by a crisis in Russian society's relationship with nature—shortages, famine, and disease—focused Vernadsky's attention on the connections between living matter—including humans—and the non-living matter of Earth.[26] As a geochemist, his career before World War I had been centered on the study of non-living matter and radioactivity. His work in this area won him election to the Imperial Russian Academy of Sciences in 1912 and an international reputation.[27] But during World War I his thinking developed in a new direction. He began an analysis of the relationship between living matter and the rest of nature. As an advisor to the Russian Ministry of Agriculture in 1916, he had already become impressed by the important role living matter played in the geochemistry of Earth. But the key concept in his thinking—that of the biosphere—developed in 1917. As he expressed it in his autobiography (published in the West during the 1960s by his historian son, George), his pre-World War I experience in geochemistry and his work during the war effort "opened a whole new world for me. I realized that the basis of geology lies in the chemical element—in the atom—and that living organisms play a prominent role, perhaps the leading one, in our natural environment—the biosphere."[28]

As discussed above, Vernadsky's doctor advised him to move South to effect a cure for tuberculosis. Vernadsky arranged to spend some time at the Academy of Sciences' Biological Research Station at Staroselskii in the Ukraine. Taking long walks in the forests and meadows of this region, he was struck for the first time by the enormous speed with which living organisms can multiply and their power to change the natural environment in a short time. He noted in particular a phenomenon the peasants in the region called "the flowering of the waters," which he realized was caused by the rapid multiplication of algae in the local ponds and wells. This phenomenon occurred virtually overnight with the favorable chemical conditions brought about by the arrival of warm weather.

The rapid multiplication of small organisms had been known to Vernadsky only from descriptions in books. Now he saw it with his own eyes and observed its startling power to change an environment.[29] He began to think about the connections between living matter and the rest of the physical environment and to explore the significance of living matter for geology and geochemistry. In the process of studying mineralogy, he had already seen the enormous influence of living organisms in the history of chemical elements. That is, he realized that one could not understand geology without looking at the influence of life on non-living matter. By 1922, he could write that of ninety-two elements then known, between fifty and sixty were closely tied with the history of living organisms. It was precisely these elements that composed 99.6 percent of the weight of the whole earth's crust.[30]

As one of his Russian biographers has noted: "Before Vernadsky the major Austrian geologist E. Suess had introduced into science the idea of the biosphere as the particular covering of the earth's crust which contained life. But now Vernadsky saw in the biosphere the most characteristic feature of the mechanism of our planet, the thick covering containing living matter in which is concentrated free chemical energy created out of the energy of the sun."[31] Here his habits of mind—thinking both as a geologist and a chemist—were applied to biology. He began to think of all living matter as a kind of geological layer or stratum, in terms of its size, shape, extent, mass, and historical evolution. Rather than look at life from the standpoint of *forms* of individual organisms and species (morphology), as biologists tended to do, he looked at it from the viewpoint of the overall mass, chemical composition, and energy of "living matter" in the biosphere.[32] Biogeochemistry, therefore, was based on the dual concepts of matter and energy. The biosphere consisted of physico-chemical organisms, living and non-living. Each organism could be considered a field in which specific atoms existed in specific quantities, and species itself became an expression of these phenomena in numbers.[33] Over time, organisms within the biosphere maintained dynamic equilibrium through cycles of decay and transfer, carried out chemically through the migration of atoms among and between living and non-living matter.

Vernadsky considered the weeks spent in Staroselskii some of the most creative of his life. Moving to his dacha at Shishaki in the Ukraine, he began to express his new ideas on paper. As the rest of the nation was caught up in the revolutionary chaos that summer of 1917, Vernadsky was infected with enthusiasm for his new ideas. He wrote as he lay in the tall grass of his native Ukraine for long hours, not even noticing the mosquitoes that bit him.

> There, in the meadows, I worked at a fast tempo. I clarified for myself the basic concepts of biogeochemistry, sharply differentiated the biosphere from the other envelopes of the earth and realized the basic significance of the multiplication of living matter in the biosphere.
>
> I began to write with great excitement, with a broad plan of explanation. It seems to me now that that simple and new concept of living matter as the totality of living organisms which I introduced into geochemistry saved me from those complications which imbue contemporary biology, where life is presented basically as the opposite of inorganic matter.[34]

Instead, as a geochemist, Vernadsky began to look at the totality of life as a form of matter inseparably connected with the rest of nature:

> The concept of "life" is inextricably linked with philosophical and religious conceptions from which biologists have not been spared. Shunting to the side the concept "life" I tried to remain on a precise empirical foundation and introduced into geochemistry the concept of "living matter" as the totality of living organisms, inextricably linked with the biosphere as an inseparable part of it and function of it.
>
> Since that time, wherever I found myself and under whatever circumstances (sometimes quite difficult), I worked constantly reading and thinking about problems of geochemistry and biogeochemistry and I continue to work in this manner now.[35]

One of his Soviet biographers later commented that "what he wrote in Shishaki on forty pages of graph paper was the basis of a new science, a 'new world view,' revolutionizing the traditional thinking, not only of geologists."[36] His ideas were resisted for some time by Soviet biologists, who were working from a different paradigm and some of whom viewed Vernadsky as an interloper. Other colleagues within the Soviet scientific establishment were also unhappy with his new direction. According to his most famous pupil, A. P. Vinogradov, Vernadsky's new interest was greeted with derision by some of his associates, who felt he was abandoning a study of the earth's chemistry for the "geochemistry of the soul of the mosquito."[37]

In the chaotic social environment of the next six years, Vernadsky worked on his new ideas. Whether in Kiev, roaming about the Ukraine seeking support from the various governments which came and went during the civil war, recovering from a nearly fatal case of typhus in 1919, or in Paris teaching at the Sorbonne in the early 1920s and working with the Curies at their Radium Institute, he gradually developed the foundations for what he called bio-geochemistry. As he wrote to his wife once during these troubled years: "It is strange to look on myself and at the whole course of history with all its tragedies and personal experiences from the point of view of a passionless chemical process of nature."[38]

Vernadsky began to publish his first works on biogeochemistry even before leaving Soviet Russia in 1922, and his years in the West saw a continuing series of studies on aspects of his new interest, both in Russian and in French. The leading French scientific journal, *Revue Générale des Sciences,* as well as the *Proceedings of the French Academy of Sciences,* published a number of these papers.[39]

As early as 1923, Vernadsky drafted a lengthy memorandum proposing the establishment of a laboratory that would pursue the study of the chemical relationships between living and inert matter. He sent this memo during 1923 to the secretary of the British Association for the Advancement of Science, the National Research Council in the United States, and the Carnegie Institution of Washington, D.C.[40] None of these sources was encouraging about the prospects for funding. His friend the American mineralogist E. S. Dana of Yale, who had sent him money during the civil war, indicated that Vernadsky had contacted him about moving permanently to the United States in order to create a biogeochemical lab there.[41] (Vernadsky had already confided to one of his Russian pupils his belief that the center of scientific activity in his major areas of interest was moving from Europe to the United States.)[42]

But the American scientific community rejected his proposal. Dana, the editor of the *American Journal of Science,* wrote him on April 3, 1923:

> I can easily appreciate, now that your entire family is out of Russia, how much you would regret returning to that country so long as conditions are what they are at present. I wish that we could help you to some position in America which would afford you the support that you need. At the moment, however, I am unable to suggest anything, notwithstanding the position that you occupy in the scientific world and the notable work which you have accomplished.[43]

His friend the American historian F. A. Golder, who had worked with the American Relief Administration in Russia, passed Vernadsky's proposal to scientists at Stanford, where Golder was then teaching. In a letter of January 15, 1924, Golder wrote to Vernadsky:

> I agree with you that there is not much chance for scientific work in Russia and I wish you had better opportunities elsewhere. The reports that reach me from time to time are not favorable. I had hoped that our government would recognize the Soviet and make it easier to extend help to those who need it.[44]

On February 12, 1924, the Stanford scientist Robert E. Swain, then chairman of the Chemistry Department, replied to Vernadsky's proposal in a discouraging way:

> I have read with some interest these two papers from Dr. Vernadsky. My feeling is that while he is quite right in his estimate of the importance of organized work along the lines mentioned, it hardly deserves the attention he would give it. As a matter of fact, a considerable amount of work is now in progress in this country as well as abroad along just these lines. I sometimes think that we have too many organizations in the scientific world to claim our attention and time.[45]

Swain did not mention who was pursuing the same line of research; in fact it seems doubtful that anyone was, and Swain probably meant this as a polite way of turning Vernadsky down. One of the latter's difficulties was that his best Western contacts at this time were primarily mineralogists and geologists who knew his pre-revolutionary reputation for work in these areas. The chemists and biologists whose support he needed to become established in his new field of interest did not know him and generally did not share his holistic approach. In addition, they may well have been skeptical of an endeavor by a crystallographer and geochemist who was moving so late in his career (Vernadsky was sixty in 1923) to a new, rather speculative area on the frontiers of scientific knowledge.

Vernadsky also tried during 1924 to remain permanently in France, creating his new lab as part of the Natural History Museum in Paris. But this effort also came to naught, Vernadsky indicated, due to French financial problems.[46]

Throughout this period, Vernadsky kept up his ties with the Soviet Academy, receiving reports from his assistants at the Radium Institute and KEPS, both of which he still headed. They also helped by sending him extensive bibliographic and other research materials he requested for his work in biogeochemistry.[47] In Paris and Prague during 1925 he finished his book *The Biosphere*, which was published in Leningrad in 1926 and in a French edition in 1929.[48]

Whether or not the Soviet authorities were aware of his efforts to remain in the West is unknown, but in 1925 they tried to encourage his return by awarding him a newly created chair in the Academy, with the freedom to pursue his own line of research, for which he had been unable to find adequate funding in the West.[49]

The Soviet government had a keen appreciation both for Vernadsky's enormous scientific prestige and his moral authority. As one of fewer than a hundred members of the old Imperial Academy of Sciences, he would, by his return and cooperation with the Soviet government, add legitimacy to the new regime and help to attract the old intelligentsia whose experience and expertise were desperately needed for Soviet reconstruction efforts. While Vernadsky's age and new interests worked against him in the West, in the Soviet Union he could command enormous respect and greater material resources, especially now that the Soviet economy had begun to recover. Beyond that, he hoped to attract bright, enthusiastic young people who would carry on his efforts once he had departed the scene.[50]

On March 7, 1925, Vernadsky wrote to a close friend from Paris, the Czech professor F. Slavik:

> I am deeply confident of the great future—not far off—of biogeochemical research work for practical life and for the development of human thought, for the success of other sciences, in particular biology, geology, chemistry, and mineralogy. This development is impossible or will go very slowly without the creation of a scientific center. There is no such center now. Both in Washington (Clarke) and in Christiana, Norway (Goldschmidt), in both instances no attention is devoted to the phenomena of life, and this is impossible for the attainment of exact results.[51]

Vernadsky was in occasional correspondence with F. W. Clarke and was on closer terms with V. M. Goldschmidt, as indicated by the extensive correspondence between the two men.[52] In 1929, when Vernadsky's new lab in the USSR was already well under way, Goldschmidt wrote that he was very pleased with Vernadsky's biogeochemical research, which "has the greatest significance for geochemistry."[53] By 1933, Goldschmidt's own work reflected an interest in biogeochemistry, particularly his valuable studies of the carbon cycle in the biosphere.[54]

In the last two decades of his life, Vernadsky published several editions of his fundamental monograph, *Geochemistry,* as well as editions of *The Biosphere* in Russian and French, several editions of his *Biogeochemical Problems,* and dozens of scientific papers in these areas.[55] At the time of his death, he had completed all but the editing and the final chapter of *The Chemical Structure of the Earth's Biosphere and Its Surroundings.*[56] Vernadsky's main ideas, developed first in *The Biosphere* during the 1920s, were revised and summed up at the end of his life in *The Chemical Structure.*

The Biosphere should be viewed as an initial effort to define the problem area, to provide a general direction for future research, and to give some rather tentative early conclusions. The object of the book, Vernadsky wrote, was "to draw the attention of naturalists, geologists, and above all biologists to the importance of a quantitative study of the relationships between life and the chemical phenomena of the planet."[57] Approaching the definition of the biosphere as a geologist, Vernadsky wrote that the earth is divided into two classes of structures. The first are the great concentric regions called concenters and the second are subdivisions of these called envelopes or geospheres. There are three great concenters: the core of the planet, the Sima region, and the crust. The matter in each of these is unable to circulate from one concenter to another except very slowly and at certain fixed rates. The biosphere is the envelope or upper geosphere of one of these great concenters, the crust. By definition, the biosphere is the only geosphere which contains living matter. Vernadsky devoted a good deal of space in the volume to defining the limits of the biosphere, its height above the earth's surface and depth below, that is, the limits of life.[58]

One of Vernadsky's basic assumptions was that nature is not governed by accident but is ruled by laws that can be expressed, generally, in quantitative form. "Earth's structure," he wrote, "is a harmonious integration of parts that must be studied as an indivisible mechanism. . . . Creatures on earth are the fruit of a long and complicated mechanism in which it is known that fixed laws apply and chance does not exist."[59]

Vernadsky intended his book to be an attack on several prevailing notions in science: not only the approach of biologists who saw life as the study of separate organisms and species rather than the totality of "living matter." He also attacked the views of geologists who thought that the chemical composition of the crust was a result largely of terrestrial processes such as metamorphism (temperature and pressure) and who disregarded the influence of living matter in the formation of the earth's crust, as well as the influence of cosmic radiation on living matter:

> Cosmic forces from outside in large measure shape the face of the earth, and as a result, the biosphere differs hypothetically from the other parts of the planet. . . . We can gain insight into the biosphere only by considering the obvious bond that unites it to the entire cosmic mechanism. . . . The biosphere may be regarded as a region of transformers that convert cosmic radiation into active energy in electrical, chemical, mechanical, thermal and other forms. Radiations from all stars enter the biosphere, but we catch and perceive only an insignificant part of the total; this comes almost exclusively from the sun.[60]

Vernadsky saw the biosphere as chemically the most active part of the earth's crust. That chemical activity was directly traceable to the input of cosmic (especially solar) radiation and the way that radiation was transformed by living matter within the biosphere.

As for organisms and their geological influence, Vernadsky believed that organisms must be studied not only alone or even by groups such as species but in their overall mass effects. The mass of living matter at any one time, he estimated, is very small and its weight does not exceed 0.25 percent of the earth's biosphere.[61] But this matter is found in constant and ceaseless motion, in interaction with inert matter, giving birth in this way to one of the great planetary phenomena: the migration of chemical elements in the biosphere. One of the chief problems he defined for biogeochemistry was the tracing of these migrations and their influence through nature. Vernadsky observed that thanks to the "whirlwind of molecules" set in motion by life and the cosmic energy that fueled life, the chemical influence of organisms penetrates far into the surrounding environment. The masses of inert solid matter passing through living matter are much larger, he noted, than the weight of living matter existing at a given moment on the planet. In this way the biogenic migration of atoms propelled by living matter, he argued, is found at the basis of all the most important chemical transformations taking place in the earth's crust.[62] It is, for example, directly involved in the formation of the earth's atmosphere, once thought by geologists to be gases attracted to the earth by gravity. The biogenic migration of atoms also plays an integral role in the formation of mineral deposits in the earth's crust.

Vernadsky challenged natural scientists to measure, using the newest physical methods, the energy of living organisms in their different forms of activity. This, he wrote, can lead to the discovery of hitherto hidden sources of energy and more exact quantitative evaluation of the geophysical effects of life on the surrounding environment. Vernadsky was particularly concerned to reduce to a mathematical formula the laws governing the multiplication of living organisms. If not prevented by some external obstacle, he argued, each organism could cover the whole globe in a time that is different but fixed for each species. Vernadsky developed a formula for the maximal speed of multiplication of various species, with the qualification that the actual speed of transmission of life varies with external conditions, such as the intensity of solar radiation, access to food, gas to breathe, space to grow, etc.[63]

Vernadsky was particularly concerned to discover which elements were involved in life processes and in what proportions. In 1930, he published his first calculations on the chemical composition of all living matter, which were later refined by his student Vinogradov.[64] He also set up a card catalogue in his new laboratory to collect data on the average chemical composition of as many different species as possible.[65] This provided a new way of classifying species, different from morphological classification, and also was valuable to those studying the geochemical activity of the various species that make up a particular ecosystem.

A final question that Vernadsky only touched on in *The Biosphere* but then devoted considerable thought to in later years involved the differences between living and non-living (or inert) matter. In this initial volume he asked: "Are the atoms which have been absorbed by living matter the same as those of crude matter or do special isotopic mixtures exist among them? Only experiment can give an answer to this problem."[66] He went on to suggest that a profound change in atomic systems probably occurs in living matter and that living matter forms a special thermodynamic field within the biosphere. This problem was one of the central concerns of his laboratory from its founding, where the study of isotopes in living matter became a major focus of research.[67]

There were several problems in Vernadsky's initial treatise on the biosphere. For one thing, he gave no real hint of the immense difficulties involved in trying to measure the quantity of living matter in the biosphere or the actual speed of multiplication of individual species. With regard to measuring the mass of living matter, the American geochemist Brian Mason wrote in 1962:

> Here we are faced with many difficulties, difficulties that were not present in other calcuations of this kind. The matter of the biosphere is not uniformly distributed, as is, for example, that of the atmosphere. It is also in a constant state of change, and the cycle of change is very rapid. The life span of any organism is minute in comparison to geological time, and the life cycles of different organisms are immensely varied. . . .[68]

With regard to the second problem, Vernadsky himself noted at the end of his life that little progress had been made in gathering exact data on the average speed of multiplication of various species.[69]

Beyond that, Vernadsky made several assertions in *The Biosphere* that proved highly controversial and that he himself began to qualify before his death. Despite his claim to eschew speculation in favor of gathering only verified empirical generalizations in this book, some of his assertions were actually quite speculative and have been challenged by the work of other scientists.

Vernadsky asserted that the notion of a beginning of life on this planet, abiogenesis (direct creation of living matter from crude), was "illusory, harmful, and even dangerous when applied to contemporary geology." Living matter has probably always existed in the universe, he suggested. Equally harmful, he believed, was the view that there were pre-geological stages of planetary evolution during which conditions on earth were clearly different from those which can now be studied.[70] During all geological periods, he argued, there has never been any trace of abiogenesis. Throughout geological time, no periods devoid of some form of life have ever been found. The conditions of the terrestrial environment during all this time have favored the existence of living matter, and conditions have always been approximately what they are today. This was in large part, Vernadsky believed, because the amount of radiation reaching Earth from the sun remained constant throughout geological time. Therefore, in all geological periods the chemical influence of living matter on the surrounding environment has not changed significantly. The average chemical composition of both living matter and the earth's crust have been approximately the same as they are today.

All these arguments, Vernadsky claimed, were backed by empirical evidence, whereas, in fact, they were very controversial. Vernadsky saw great evolutionary change in the *forms* of living matter through time, but he saw little change in the terrestrial environment or in the quantity, composition, and influence of "living matter" throughout geological time.

In the same period that Vernadsky was putting forth such views another Soviet scientist, A. I. Oparin, was developing his hypothesis concerning the origins of life from non-living matter at a pre-geological stage in the earth's development, a view that has gained wider acceptance than Vernadsky's.[71] Among contemporary scientists, a different view about the physical conditions of the earth's non-living environment has gained credence. Many scientists now think that during the earth's existence, the sun's output of energy has increased substantially, rather than remaining constant, as Vernadsky thought. According to one of these sources, "the earth now received between 1.4 and 3.3 times more energy [from the sun] than it did just after its formation 4,000 million years ago."[72] At least in part as a result of this, the terrestrial environment has changed radically: "During the period that life has existed on earth, at least 3 giga-years, there have been profound changes in the chemical and physical environment."[73]

Vernadsky himself never held his views as dogma, and by the early 1940s, when he wrote *The Chemical Structure of the Earth's Biosphere,* his views on the origins of life and the unchanging nature of the earth's physical environment became somewhat more cautious and qualified. He simply stressed the lack of evidence of abiogenesis during *known* geological periods (eras which scientists had been able to study) but admitted that abiogenesis might have taken place at some earlier point.[74] He continued to stress that physical conditions had varied little during this period of perhaps two billion years, although today he clearly seems wrong in this emphasis. Nonetheless, Vernadsky was modest in his conclusions. Even in *The Biosphere* he emphasized the limits of present knowledge: "We do not know how the extraordinary mechanism of the earth's crust could have been formed or how it became saturated with life, nor do we know the origin of the matter of the biosphere, which has been functioning for billions of years. It is a mystery, just as life itself is a mystery in the framework of our knowledge."[75] By the time he wrote *The Chemical Structure,* he was willing to believe that this mystery was, in principle, solvable.[76]

In his final book Vernadsky was able to do two things: to provide some answers for questions he had posed in 1926 and to broach some new and important problems. In *The Biosphere* he asked if the isotopic composition of living matter is different from inert matter. By the early 1940s he was able to answer this question in the affirmative, based on the extensive work of his lab.[77] The evidence also indicated, Vernadsky wrote, that radioactive elements are found in all organisms. For each species, the amount is constant, different, and characteristic of that group.[78] On the differences between living and non-living matter, with which he had also been concerned in the 1926 volume, Vernadsky considered his work in this area to be "my last creative contribution to science."[79] In fact, he was continuing the work of two famous predecessors in this area, Louis Pasteur and Pierre Curie, as he freely admitted. According to Pasteur, living and inert matter differed spatially and geometrically. In particular, Pasteur had discovered the arrangement of organic molecules to be disymmetrical, unlike the symmetrical geometry of crystals in non-living matter. Vernadsky suggested that the disymmetry discovered by Pasteur was connected with atomic structure and the fields occupied by atomic particles. The structure of molecules in living beings, Vernadsky wrote, possesses elements of symmetry that are not possible in hard crystal spaces. Molecules in living matter—all its proteins, fats, and carbohydrates—are not subordinate to the laws of symmetry of inert bodies in the sense that all of them essential to life are sterically left. Vernadsky asked himself the reason for this. As early as 1938, he put forward the thesis that the geometry of living organisms is not Euclidean but Riemannian. Vernadsky developed this idea in his final book.[80]

Vernadsky's thoughts on this problem were clearly speculative, more suggestive than definitive, but he had raised an issue—the geometry of matter in the universe and particularly the significance of disymmetry for living matter—that has become important in more recent years.[81] Clearly, if the atoms in living matter and their arrangement were different from those in inert matter, these differences were crucial to an understanding of the laws governing the relationship between the living and non-living matter of the biosphere. It was this problem that Vernadsky posed in his final work for other scientists to pursue after him.

In this final work, Vernadsky summarized what he had learned concerning the chemical functions of life, including the gas functions, the role of organisms in concentrating particular minerals and forming deposits of these in the biosphere, and the geochemical functions of mankind. Vernadsky failed to complete more than a sketch on the latter subject before his death.[82] This lacuna is unfortunate, since he clearly believed that humanity was becoming the dominant form of living matter in the biosphere and was transforming it into something essentially new: what he called the noosphere (borrowing the term from Le Roy and de Chardin, but giving it a more precise meaning).[83] By noosphere he meant a geosphere dominated by human reason and conscious work activity, which were rapidly changing the chemical structure of the biosphere. His final work provides only a few suggestive passages on a subject which clearly occupied his thoughts in the years immediately preceding his death. For example, Vernadsky noted that since the eighteenth century the quantity of biogenic gases produced by humans had increased greatly: "Humans cutting down forests and fields . . . change the face of the planet, create numberless new physical-chemical processes in the history of the biosphere, until now acting more or less unconsciously. In the noosphere, the regulating of this function of humans must be one of the basic features of its new structure."[84]

Beyond such suggestive passages, which have provided useful quotes for contemporary Soviet ecologists, Vernadsky left little specific information on the biogeochemical role of human beings. The sand in his hourglass had run its course; it was left to his students, including Vinogradov, Fersman, and others, to pursue the implications of the problems he had helped to formulate.[85] A. P. Vinogradov, subsequently vice-president of the Soviet Academy of Sciences, conducted biogeochemical research into the process of photosynthesis and attempted to determine the laws governing the dispersion of chemical elements in the upper layers of the earth's crust. Aleksandr Fersman continued an active career researching and teaching geochemistry. He studied the behavior of chemical elements under various thermodynamic, physical, and chemical conditions and introduced the concept of "clarkes of elemental concentration"—the frequency of distribution of chemical elements in rocks according to percentage of atomic weight. Perhaps as a brainchild of Vernadsky's concept of the noosphere, Fersman proposed the concept of "technogenesis"—the effect of economic and industrial activity on the chemical elements in the earth's crust or, more simply, the extraction and redistribution of chemical elements caused by agriculture and engineering. Other students were instrumental in the geological and geochemical sciences, where they pioneered the production of artificial minerals, the location and distribution of rare metals within the USSR, and the use of mineral fertilizers.

While some of Vernadsky's ideas have been revised by later researchers, many of his formulations have withstood the test of time and have become a part of world science.[86] In fact, many of Vernadsky's ideas are so well known today that they often seem self-evident and perhaps even elementary. It is easy to forget their freshness and originality in his own time. For example, the biogeochemical principle he formulated concerning the direction of evolution seems commonplace today. This principle states that the "evolution of species during geological time goes in the direction of increasing the biogenic migration of atoms in the biosphere."[87] Those species which have survived and increased in numbers are those that have been able to increase the biogenic migration of atoms. Evolution, therefore, has a definite direction, which Vernadsky connected with the growth of central nervous systems and with the capability of those species possessing the most highly developed consciousness to increase their biogeochemical activity and thereby the migration of atoms in the biosphere. Here Vernadsky pursued the ideas of the nineteenth-century American zoologist D. Dana concerning the development of central nervous systems during the course of evolution. In so doing, his directional view of evolution was a revision of Darwin.

From this starting point, Vernadsky made a leap of faith into the future. He went beyond an empirical knowledge of the past and present to view the process of evolution optimistically, as necessarily leading to the final triumph of humanity and human reason, its ability to shape the future environment for human benefit.[88] This optimistic leap of faith largely blinded him to another possible alternative future: that of ecological disaster, which has concerned so many of his successors in the West.

Vernadsky devoted his energy to developing a scientific concept of the biosphere to satisfy his intellectual curiosity and create a coherent view of nature, that is, to satisfy a traditional concern of many Russian intellectuals for monistic knowledge that would make the universe comprehensible and provide an explanation for all its phenomena. Unlike many Russian Marxists and others who were attracted to monistic philosophies, Vernadsky, as a scientific liberal, was willing to live with uncertainties and to accept the limitations and tentativeness of all knowledge. It was the search for scientific truth, rather than the certainty of having attained it, that seems to have fascinated him and provided a purpose to his life.[89] But Vernadsky also sought a *raison d'être* for his work in the practical applicability of scientific knowledge for human betterment. While he had an enormous curiosity about the universe and pursued his interest in the biosphere in part because it was an intellectual activity that satisfied some of his curiosity, he justified his work particularly in terms of its potential usefulness. He perceived its practical applications primarily as a stimulus to economic development, for example, locating useful mineral deposits in the biosphere, creating artificial mineral resources, and increasing agricultural productivity.[90]

The perils of industrialization and the usefulness of biogeochemical knowledge in helping to overcome those perils were not central concerns in Vernadsky's work. This is what makes him a product of his own time and not of the more recent era with its concern for the use of ecology to solve environmental problems. Except for a brief period before and immediately after the revolution during which Vernadsky helped to form one of the first Russian preserves for scientific study of rare mineral deposits, his career shows little involvement in environmental movements.[91] His concern was expressed only in a limited sense:

> The face of the planet—the biosphere—is being sharply changed chemically by man both consciously and even more so unconsciously. Man is changing physically and chemically the atmosphere and all the waters of nature. As a result of the growth of human culture in the twentieth century, the shores of the sea and part of the ocean are changing more and more radically (chemically and biologically). Man must now take greater measures in order to conserve for future generations the riches of the oceans which belong to no one.[92]

Where Vernadsky's thinking might have led if he had lived to see the increase in environmental problems after 1945 is anyone's guess. The fact remains that his work showed little active concern for conservation until 1944; nor did he indicate until then an awareness that economic development and sophisticated technology might have dangerous side effects. Vernadsky's legacy for Soviet conservationists is ambiguous. His optimistic, promethean view of humanity reinforces dominant Soviet attitudes toward the economic exploitation of the environment and creates problems for "preservationists" within the conservation movement. But at the same time, he and his students provided many of the tools of geochemical analysis that are vital to environmentalists in demonstrating the possible detrimental long-range consequences of economic activity and providing a new, more careful approach to economic development. As one Soviet scholar interprets his work, Vernadsky "wanted science to find the optimal interrelationships of chemical exchange between society and nature, in order to satisfy the demands of society for the products of nature while conserving and preserving natural resources, not destroying the organization and stability of the system 'nature-society.' "[93] While this scientist may exaggerate Vernadsky's concern for the environment, Vernadsky's thought is consistent with the "wise use" position, if not the preservationist strain, within twentieth-century environmental movements.

Beyond that, Vernadsky left not only a great deal of specific knowledge about the biosphere and humanity's place in it but also a number of general statements useful to Soviet conservationists, such as the following:

> In the intensity, complexity, and depth of modern life, man forgets in a practical sense that he himself and all humanity, from which he may not be separated, is inescapably linked with the biosphere. . . . In reality no one living organism finds itself in a free circumstance on Earth. All these organisms are constantly and inextricably linked—first of all in their food and breathing—with the material-energy environment around them. Outside of it they cannot exist under natural conditions.[94]

Perhaps even more valuable than the specific knowledge of the biosphere, Vernadsky left a world view in which knowledge is derived from the free clash of scientific schools and is revised to fit new evidence, freed of dogmatic outside influences. As another Soviet scholar has expressed it, "It is necessary to remember that V. I. Vernadsky never gave to his scientific conclusions the character of dogma and always held in view the relative character of scientific truths, relative to the knowledge of a particular time."[95]

This view and Vernadsky's assertion of the primary value of freedom of thought for science and for human creativity remain his greatest legacies for Soviet culture and world science. Despite the collapse of the liberal political party which he helped found in Imperial Russia, Vernadsky remained true to a liberal ideal, trying to reshape that ideal so that it might survive under Soviet conditions. Whether this ideal will be realized as Vernadsky hoped only the future can tell. His brand of scientific liberalism, maintained in more recent times by Academician Andrei Sakharov and others, faces what must seem like overwhelming odds. That Vernadsky's scientific ideal remains a living part of culture in the Soviet Union, however, is indicated by the revival of interest in his work and the publication there of his hitherto unpublished manuscripts.

Beyond that, Vernadsky personifies an important tendency within the Russian intelligentsia to pursue science as a major alternative to both political dogma and traditional Russian Orthodoxy. This belief provides a bridge for many Russian intellectuals between the old society and the new. Some of those who, like Vernadsky, disliked Bolshevism were able to persuade themselves that the intellectual appeal of Marxism-Leninism was ephemeral and that they could pour the new wine of science, secular culture, and economic development into the old wineskins of a moldy Marxist movement. Be this illusion or self-delusion, we need to recognize the force of this tendency among members of the non-party intelligentsia.

Vernadsky's significance for twentieth-century science probably rests more in his ability to frame new questions than in the specific biogeochemical data his work contains, much of which are dated, or in his conclusions, some of which are commonplace today and others considered mistaken or misleading. Despite his desire to be known as someone who kept close to empirical reality and framed generalizations based on scientific fact, it is perhaps his speculative genius that forms his strongest legacy for world science. As G. E. Hutchinson expressed it in a recent letter, "Vernadsky's importance does not lie in the factual content of any of his writings but his having established what I can only call a *speculative tradition* which has proved to have immense stimulatory power."[96]

Vernadsky may have realized this. In his final work, he wrote that a scientist "cannot forget, as Newton clearly expressed at the end of his life, that compared with the masses of questions which we can pose and attempt to answer about the awesome phenomena of reality, he was like a little boy constructing sand castles on a beach."[97]

Time will pass, Vernadsky seemed to be saying, and the wind, the waves, and the feet of those who come later will wear down the most elaborate structures. Others will build new ones, perhaps solving problems which their predecessors posed, but will "again feel like small boys confronted with new questions which have arisen before them." Vernadsky sensed the fragility of his concepts in the presence of eternity and the forces of nature, but he hoped that his example would stimulate others to explore questions to which he had devoted a lifetime.

Notes

1. A LIFE FOR SCIENCE

1. V. I. Vernadsky to George Vernadsky, October 6, 1944, Vernadsky Papers, Box 25, Folder Vernadsky-177, Bakhmeteff Archive of Russian History and Culture (hereafter BAR), Columbia University, Special Collections.

2. On the Nihilists, see Daniel Brower, *Training the Nihilists: Education and Radicalism in Tsarist Russia*, Ithaca, New York, 1975; Alain Besançon, *Education et société en Russie dans le deuxième tiers du XIXe siècle*, Paris, 1974.

3. See Marc Raeff, *Origins of the Russian Intelligentsia: The Eighteenth Century Nobility*, New York, 1966 and Walter M. Pinter and Don K. Rowney, eds., *Russian Officialdom: The Bureaucratization of Russian Society from the Seventeenth to the Twentieth Century*, Chapel Hill, North Carolina, 1980.

4. "O rode Vernadskikh," unpublished manuscript by George Vernadsky, dated 29 July 1936, in Vernadsky Papers, Box 4721, BAR, p. 1. Also see Vladimir Vernadsky's short autobiography published in *The Annals of the Ukrainian Academy of Arts and Sciences in the U.S.*, vol. XI, no. 1–2 (31–32), 1964–68, pp. 3–31.

5. "O rode Vernadskikh," p. 3; Lev Gumilevskii, *Vernadskii*, Moscow, 1967, p. 10; V. L. Lichkov, *Vladimir I. Vernadskii*, Moscow, 1948, pp. 8–10.

6. On the nature of the eighteenth-century Russian Orthodox clergy, see Gregory L. Freeze, *The Russian Levites: Parish Clergy in the Eighteenth Century*, Cambridge, Mass., 1977.

7. G. Vernadsky, "O rode Vernadskikh, p. 3; V. Vernadsky, autobiography, p. 5.

8. V. Vernadsky, autobiography, p. 4; Gumilevskii, p. 11.

9. Letter to N. E. Staritskaia, 29 May 1886, Arkhiv Akademii nauk (hereafter AAN), f. 518, op. 7, ed. khr. 33, 1. 6, quoted in I. I. Mochalov, *V. I. Vernadskii—chelovek i myslitel'*, Moscow, 1970, p. 74.

10. G. Vernadsky, "O rode Vernadskikh," p. 3; letter of V. I. Vernadsky to P. L. Dravert, 16 May 1944, in Rukopisnyi otdel gosudarstvennogo muzeia Omskoi oblasti, quoted in Mochalov, p. 74.

11. V. I. Vernadsky to S. V. Korolenko, 3 August 1942, AAN. f. 518, op. 2, ed. khr. 52, 1. 3, quoted in Mochalov, p. 48.

12. G. Vernadsky, "O rode Vernadskikh," p. 3.

13. See D. A. Miliutin, *Istorii voiny Rossii s Frantsiei v 1799*, vol. 4, St. Petersburg, 1853, p. 303.

14. V. G. Korolenko, *Sobranie sochinenii*, vol. 6, Moscow, 1954, p. 112.

15. G. Vernadsky, "O rode Vernadskikh," pp. 4–6.

16. A. D. Shakhovskaia, *Kabinet-muzei V. I. Vernadskogo*, Moscow, 1959, p. 9; Gumilevskii, p. 11.

17. G. Vernadsky, "O rode Vernadskikh," pp. 6–7; S. O. Seropolko, "Vernadskaia," *Zhenskoe delo*, 10 October 1910, pp. 3–4.

18. The Vernadsky journal was entitled, in Russian, *Ekonomicheskii ukazatel'* and was published from 1857 until 1864. See Gumilevskii, p. 12.

19. Richard Stites, *The Women's Liberation Movement in Russia*, Princeton, 1978, pp. 35–36.

20. Gumilevskii, p. 12.

21. See, for example, N. G. Chernyshevskii, *Perepiska Chernyshevskogo s Nekrasovym, Bogoliubovym, i Zelenym, 1855–1862*. Moscow, 1925.

22. Aleksandr V. Nikitenko, *Moia povest' o samom sebe i o tom chemu svidetel' v zhizni byl. Zapiski i dnevnik (1804–1877 gg)*, St. Petersburg, 1904–5.

23. Ibid.

24. Gumilevskii, p. 14.

25. G. Vernadsky, "O rode Vernadskikh," p. 8; *Rodoslovnaia kniga Konstantinovichei*, Kiev, 1909.

26. G. Vernadsky, "O rode Vernadskikh," pp. 9–10; Gumilevskii, p. 15.

27. G. Vernadsky, "O rode Vernadskikh," p. 10.

28. AAN, f. 518, op. 2, no. 36, khronologiia, 1868, p. 1; Gumilevskii, p. 16; Shakhovskaia, p. 20.

29. AAN, f. 518, op. 2, no. 30. khronologiia, 1874, pp. 1, 4, 5; Shakhovskaia, p. 20.

30. V. I. Vernadskii to N. E. Staritskaia, 21 June 1886, AAN f. 518, op. 7, ed. khr. 33, 1. 21, quoted in Mochalov, p. 9.

31. AAN, f. 518. op. 2, no. 30, khr. 1880, pp. 1–2; G. Vernadsky, "O rode Vernadskikh," p. 5.

32. V. I. Vernadskii to N. E. Staritskaia, 6 June 1886, cited in V. S. Neapolitanskaia, "Iz vyskazyvanii V. I. Vernadskogo," in *Zhizn' i tvorchestvo V. I. Vernadskogo po vospominaniiam sovremennikov (Ocherki po istorii geologicheskikh znanii*, vypusk 11, Moscow, 1963.

33. Gumilevskii, p. 17.

34. Ibid., p. 19.

35. AAN, f. 518, op. 2, "Iz vospominanii"; G. Vernadsky, "O rode Vernadskikh," pp. 7–8; Gumilevskii, p. 19; Lichkov, p. 8; Vladimir Vernadsky, autobiography, 1924, p. 16.

36. Ibid; AAN, f. 518, op. 2, no. 46, khr. 1924, p. 16.

37. Shakhovskaia, pp. 20–21.

38. Patrick P. Dunn, "Childhood in Imperial Russia," in Lloyd de Mause, ed., *The History of Childhood*, New York, 1974, pp. 383–406.

39. For background on the classical gymnasia, see Allen Sinel, *The Classroom and the Chancellery: State Educational Reform in Russia under Count Dmitry Tolstoy*, Cambridge, Mass. 1973, pp. 130–213.

40. V. I. Vernadskii, "Iz proshlogo (Otryvki iz vospominanii ob A: N. Krasnove)," *Ocherki i rechi*, vol. 2, Petrograd, 1922, p. 101.

41. Ibid.; Lichkov, p. 11.

42. Ibid., pp. 102–103; AAN, f. 518, op. 3, Pis'ma A. N. Krasnova; Shakhovskaia, p. 21; Gumilevskii, pp. 20–22.

43. V. Vernadskii, "Iz proshlogo," p. 102; *Zhizn' i tvorchestva V. I. Vernadskogo*, 1963, p. 56.

44. Ibid.

45. Ibid.

46. Ibid., p. 103; AAN, f. 518, op. 1, no. 132, "Tetrad' s zametkami," 1877.

47. Shakhovskaia, p. 21; Gumilevskii, p. 23.

48. AAN, f. 518, op. 2, no. 24, Dnevnik, 1944, p. 7; Shakhovskaia, p. 22.

49. AAN, f. 518, op. 2, ed. khr, 1881, 1. 6; Gumilevskii, p. 24.

50. James McClelland, *Autocrats and Academics: Education, Culture and Society in Tsarist Russia*, Chicago, 1979, pp. 95–106.

51. Samuel Kassow, "The Russian University in Crisis, 1899–1911," Ph. D. dissertation, Princeton University, 1976, pp. 35–36; Alexander Vucinich, *Science in Russian Culture*, vol. 2, Stanford, 1970, p. 186.

52. Kassow, pp. 38–40; McClelland.

53. Kassow, p. 44.

54. V. Vernadskii, "Iz proshlogo," p. 104; see also AAN, f. 518, op. 2, khr. 1881, 1. 1.

55. *Almanakh sovremennykh russkikh gosudarstvennykh deiatelei*, St. Petersburg, 1897.

56. Kassow, p. 30.

57. I. M. Grevs, "V gody iunosti," *Byloe* 12 (June 1918): 50.

58. V. A. Posse, *Moi zhiznennyi put'. Dorevoliutsionnyi period (1864–1917 gg.),* Moscow, 1929, p. 35.

59. Ibid., p. 32.

60. German Smirnov, *Mendeleev,* Moscow, 1974, pp. 173–78; O. N. Pisarzhevskii, *Dmitrii Ivanovich Mendeleev 1834–1907,* Moscow, 1959, pp. 296–306.

61. V. Vernadskii, "Iz proshlogo," p. 104.

62. Gumilevskii, pp. 29–30.

63. Ibid.

64. On Dokuchaev's career, see G. F. Kir'ianov, *Vasilii Vasil'evich Dokuchaev,* Moscow, 1966, and Alexander Vucinich, *Science in Russian Culture,* Stanford, 1970, pp. 406–10.

65. *Nauchnoe nasledstvo,* vol. 2, Moscow, 1951, pp. 745ff.

66. V. Vernadskii, "Stranitsa iz istorii pochvovedeniia," 1904, p. 87ff., and V. Vernadskii, "Iz proshlogo," 1916, p. 108, both essays republished in *Ocherki i rechi,* vol. 2, 1922.

67. P. V. Ototskii, "Zhizn' V. V. Dokuchaeva," in special issue of *Pochvovedenie,* dedicated to Dokuchaev, 1904, pp. 9–10.

68. Russkii chernozem, St. Petersburg, 1883.

69. Biogeokhimicheskie ocherki, Moscow, 1940, p. 6.

70. Lichkov, p. 15; selections from Vernadsky's "Dnevnik," AAN, f. 518, op. 1, ed. khr. 212; AAN, f. 518, op. 2, ed. khr. 4, published in *Priroda* 10 (1967): 98ff.

71. Shakhovskaia, p. 23.

72. Gumilevskii, p. 34.

73. V. V. Dokuchaev, "Pis'ma k V. I. Vernadsksomu," in *Nauchnoe nasledstvo,* Moscow, 1951.

74. See Dokuchaev's article in P. R. Ferkhmin, *Materialy po otsenke zemel' Nizhegorodskoi gubernii,* vypusk 8, 1885.

75. Published in *Trudy Vol'nogo ekonomicheskogo obshchestva,* 3 (1889): 22–29, 84–85; AAN, f. 518, op. 2, khr. 1889, 1. 22; also see Vernadsky's longer article in the same journal, "O fosforitakh Smolenskoi gubernii," 11 (1888): 263–94.

76. Lichkov, p. 15; Shakhovskaia, p. 23; Gumilevskii, p. 35.

77. Kassow, pp. 95–98; McClelland, pp. 95–105.

78. Ibid.

79. On this circle, see Grevs, op. cit., pp. 42 and 64; A. A. Kornilov, "Vospominaniia o iunosti Fedora F. Oldenburga," *Russkaia mysl',* August 1916, pp. 53–86; G. Vernadsky, "Bratstvo 'Priiutino,'" *Novyi zhurnal* (New York) 93 (1968): 148–70.

80. For example, see Lichkov, p. 14, and Kornilov, op. cit.

81. *Priroda* 12 (1967): 59.

82. Ibid., p. 100.

83. Letter to N. E. Staritskaia, 6 August 1886, AAN, f. 518, op. 7, ed. khr. 33, 1. 51, str. 11.

84. *Priroda* 10 (1967): 101–2.

85. Ibid.

86. "Dnevnikovye zapiski, 1884," AAN, f. 518, op. 2, ed. khr. 4, 11.5, 18–19, quoted in Mochalov, pp. 12–13.

87. Alston, p. 47.

88. Kornilov, op. cit.; G. Vernadsky, "Bratstvo," pp. 148–49; Grevs, op. cit.

89. AAN, f. 518, op. 2, khr. 1882, pp. 3–4; Vernadsky, "Iz proshlogo," p. 106; Kornilov, pp. 54–57; Gumilevskii, p. 27.

90. Kornilov, p. 56.

91. Ibid.; AAN, f. 518, op. 2, khr. 1884, p. 4.

92. Ibid., p. 55; "Shakhovskoi, D. I.," *Russkie Vedomosti 1863–1913, Sbornik.* Moscow, 1913, p. 198.

93. Ibid., p. 197.

94. Shakhovskaia, *Kabinet-muzei*.

95. AAN, f. 518, op. 2, khr. 1882, pp. 3–4; AAN, f. 518, op. 3, letters of 1883 to his sisters O. I. Alekseeva and E. I. Vernadskaia; Gumilevskii, p. 27.

96. Grevs, *Byloe* 16 (1921): 137–66; G. Vernadsky, "Bratstvo," p. 165.

97. V. Vernadskii, "Iz Proshlogo," p. 106.

98. Mochalov, p. 11; Kornilov, pp. 57–58. Also see Mochalov's unpublished essay on Vernadsky and Tolstoy in Vernadsky Papers, Box 31.8.6.1-I-6, Folder I-3-A-a-2-3, BAR, Columbia University.

99. G. Vernadsky, "Bratstvo," pp. 148–49, 161–62; Grevs, pp. 43–44, 54–58.

100. Kornilov, pp. 58–59; G. Vernadsky, "Bratstvo," p. 153.

101. Kornilov, p. 59.

102. V. Vernadskii, "Iz proshlogo," p. 106.

103. Kornilov, p. 63; G. Vernadsky, "Bratstvo," p. 151ff; G. Vernadsky, "The Prijutino Brotherhood," Box 32.3.4.1, Folder IV-1-80, p. 857, BAR, Columbia University.

104. Kornilov, pp. 54ff; Grevs, pp. 60–86; Posse, p. 30ff; Gumilevskii, pp. 27–28; G. Vernadsky, "Bratstvo," pp. 148–51.

105. Grevs, p. 73.

106. Kornilov, p. 58ff; Gumilevskii, pp. 36–37.

107. On Rubakin, see Alfred Senn, *A Life for Books,* Newtonville, Mass., 1977, and A. Rubakin, *N. A. Rubakin: Lotsman knizhnogo moria,* Moscow, 1967.

108. *Priroda* 10 (1967): 103.

109. Ibid.

110. Ibid.

111. Kornilov, pp. 57–58, 71–79.

112. See, for example, the photograph in Lichkov, p. 15.

113. Gumilevskii, pp. 37–39, 41.

114. Ibid.; AAN, f. 518, op. 2, khr. 1886, p. 1.

115. AAN, f. 518, op. 2, khr. 1886, p. 1; Kornilov, pp. 77–79.

116. Gumilevskii, p. 38.

117. AAN, f. 518, op. 2, khr. 1886, p. 98; Kassow, pp. 103–7.

118. On these organizations, see McClelland, p. 98; Kassow, pp. 103–7.

119. AAN, f. 518, op. 2, khr. 1887, pp. 6–7; Gumilevskii, p. 40. On Ulianov's role in the Scientific-Literary Society, see Grevs, pp. 85–86 and Gumilevskii, pp. 40–45. Also see B. S. Itenburg and A. Ia. Cherniak, *Aleksandr Ul'ianov,* Moscow, 1957.

120. Ibid.

121. Quoted in Kassow, p. 61.

122. Grevs, pp. 64–70.

123. G. Vernadsky, "Bratstvo," p. 165; Grevs *Byloe,* 1921, pp. 137–66.

124. A. Vucinich, *Science in Russian Culture,* vol. 2, p. 114; McClelland, p. 62.

125. A. E. Ivanov, in *Soviet Studies in History,* Winter 1978–79: 85.

126. Gumilevskii, p. 45; Kassow, p. 57, 108; Vydrin, *Osnovnye momenty,* p. 209.

127. McClelland, p. 16.

128. AAN, f. 518, op. 2, khr. 1887, pp. 1–2; Gumilevskii, p. 16.

129. *Stenograficheskii otchet, Gosudarstvennaia Duma,* 11 June 1908, pp. 2812–18.

130. Gumilevskii, pp. 46–47.

131. Ibid., p. 47.

132. Ibid.

133. Ibid., p. 48.

134. Ibid.; AAN, f. 518, op. 2, khr. 1887, pp. 1–2.

2. SCIENCE AND SOCIETY

1. Gumilevskii, p. 49; Mochalov, 1982, p. 83.
2. Ibid.
3. *Dictionary of Scientific Biography*, New York, 1976.
4. Ibid.
5. Ibid. p. 557.
6. Vernadsky to N. E. Vernadskaia, 1 June 1888; 21 October 1888; Vernadsky to Dokuchaev, pp. 744, 761, 765.
7. Vernadsky to N. E. Vernadskaia, 25 October 1888; 30 December 1888, pp. 774–75.
8. V. V. Dokuchaev, "Pis'ma k V. I. Vernadskomu," in *Nauchnoe nasledstvo*, Moscow, 1951.
9. Dokuchaev to Vernadsky, 3 June 1888; 30 December 1888, pp. 774–75.
10. See, for example V. I. Vernadsky to N. E. Vernadskaia, 20 June, 3 August, 4 August, 8 August, 26 August, 2 October 1888. AAN, f. 518, op. 7, ed. khr. 36; 3 January, 14 January, 24 January, 11 February, 19 February, 1 April 1888. AAN, f. 518, op. 7, ed. khr. 36.
11. Vernadsky to N. E. Vernadskaia, 29 September 1890, 15 October 1890; 8 July 1888; AAN, f. 518, op. 2, khr. 1888.
12. Mochalov, dissertation, pp. 20–22.
13. Dokuchaev correspondence, November 1889.
14. Vernadsky to N. E. Vernadskaia, see note 10.
15. Ibid.
16. Vernadsky to N. E. Vernadskaia, ibid.
17. Vernadsky to N. E. Vernadskaia, 1 January 1889, in Gol'shtein collection, BAR, Columbia University.
18. Vernadsky to N. E. Vernadskaia, 19 February 1889; 21 February 1889, 2 February 1889. AAN f. 518, op. 7, ed. khr. 36.
19. Vernadsky to N. E. Vernadskaia, 20 June 1888, in *Puti v neznaemoe*, p. 6.
20. Vernadsky to N. E. Vernadskaia, 13 June 1888, AAN f. 518, op. 7, d. 35, 1. 7.
21. Dokuchaev correspondence.
22. Ibid.
23. *Dictionary of Scientific Biography*, "V. V. Dokuchaev," New York, 1976.
24. Dokuchaev correspondence.
25. Ibid.
26. Ibid.
27. Ibid.
28. Ibid.
29. Ibid.
30. Ibid.
31. Ibid.
32. Vernadsky to N. E. Vernadskaia, 3 January, 14 January, 27 January, 1 December 1889. AAN f. 518, op. 7, ed. khr. 36.
33. Vernadsky to N. E. Vernadskaia, 1 November 1889, Gol'shtein collection, BAR, Columbia University.
34. Ibid., 6 January 1889.
35. Ibid., 1 November 1889.
36. Ibid., 23 December 1889.
37. Dokuchaev correspondence, May 1890.
38. I. M. Grevs, "Gody iunosti," *Byloe* 12 (1918); 16 (1927).
39. Gumilevskii, pp. 53–57.

40. Ibid.
41. Glavnye biograficheskie daty, 1943, Kabinet-muzei, Moscow; AAN f. 518, op. 2, d. 65 khr. 1889.
42. Gumilevskii, p. 56; Mochalov, 1982, p. 87.
43. V. M. Korsunskaia, N. M. Verzilin, V. I. Vernadskii, Moscow, 1975, p. 54.
44. See the chapter in Mochalov dissertation, "Estestvonnonauchnye osnovy mirovozzreniia V. I. Vernadskogo," "Religioznosti: ee paradoksy," Lenin Pedagogical Institute, Moscow, 1971.
45. Ibid.
46. Ibid.
47. Vernadsky to N. E. Vernadskaia, 1 December 1889, published in *Puti v neznaemoe. Pisatel' rasskazyvaet o nauke.* Sbornik shestoi, Moscow, 1966, pp. 414–15.
48. See note 41; see also Vernadsky to N. E. Vernadskaia, 26 December 1893; khronologiia za 1893, Kabinet-muzei, Moscow.
49. Vernadsky to Sergei Oldenburg, 18 January 1889, Gol'shtein collection, BAR, Columbia University.
50. Ibid.
51. Vernadsky to N. E. Vernadskaia. 20 June, 3 August, 4 August, 18 August, 26 August, 2 October 1988. AAN f. 518. op. 7, ed. khr. 35.
52. Dokuchaev correspondence.
53. Vernadsky to N. E. Vernadskaia, 15 September, 18 September 1890. AAN, f. 518, op. 7, ed. khr. 7.
54. V. I. Vernadsky, "O polimorfizme kak obshchem svoistve materii," *Uchenye zapiski Moskovskogo universiteta,* Otdelenie estestvenno-istoricheskii, 9 (1892).
55. See note 47.
56. Diary entry, 1941–43, AAN f. 518, op. 2, ed. khr. 21, 1. 82; cited in Mochalov, 1982, p. 98.
57. Vernadsky to N. E. Vernadskaia, 19 October, 1890, 11 May, 14 May 1891. AAN f. 518, op. 7, ed. khr. 37.
58. Ibid.
59. Mochalov, 1982, op. 105.
60. Vernadsky to N. E. Vernadskaia, 19 October 1891, AAN f. 518, op. 7, ed. khr. 37.
61. Ibid.
62. Kassow.
63. Ibid.
64. Ibid.
65. Kassow, p. 63; G. I. Shchetinina, *Universitety i obshchestvennoe dvizhenie,* p. 197.
66. V. I. Vernadsky to N. E. Vernadskaia, 17 April 1896.
67. Kassow, p. 89.
68. Imperatorskii Moskovskii Universitet. *Doklad komissii izbrannoi Sovetom Imperatorskogo Moskovskogo Universiteta 29ogo fevralya 1901 goda dlia vyiasneniia prichin studencheskikh volnenii i mer k uporiadocheniiu universitetskoi zhizni.* Moscow, no date, p. 11; translated in Kassow, p. 96.
69. Kassow, p. 104.
70. V. I. Vernadsky to N. E. Vernadskaia, 16 May 1896; 11 July 1896.
71. Vernadsky to Dokuchaev, 10 March 1891, Dokuchaev correspondence, p. 812.
72. V. V. Dokuchaev to V. I. Vernadsky, 15 April 1891; V. I. Vernadsky to V. V. Dokuchaev, 28 April 1891, Dokuchaev correspondence, p. 817.
73. V. V. Dokuchaev to V. I. Vernadsky, 1 May 1891, Dokuchaev correspondence, p. 818.
74. Dokuchaev correspondence, p. 852.

75. V. V. Dokuchaev to V. I. Vernadsky, 17 August 1891, Dokuchaev correspondence, p. 824.

76. V. I. Vernadsky to N. E. Vernadskaia, 28 October 1891, AAN f. 518, op. 7; f. 518, op.2, khr. 1891, 11. 2,5,6.

77. *Dictionary of Scientific Biography,* Vol. 13, New York, 1976, p. 618.

78. Balandin, 1979, p. 57.

79. Mochalov, 1981, p. 99; Henri Louis Le Chatelier, *Kremnezem i silikaty,* Leningrad, 1929; E. Scheibold, *Osnovnye idei geokhimii,* vol. 3, Leningrad, 1933; N. Bronskii, A. Raznikov, V. Iakovlev, *V. I. Vernadskii. K stoletiiu so dnia rozhdeniia.* Rostov, 1963, p. 47.

80. Gumilevskii, p. 62.

81. Balandin, 1979, p. 58.

82. *Dictionary of Scientific Biography.* vol. 13, p. 618; *O gruppe sillimanita i roli gliznozema v silikatakh,* Moscow, 1891.

83. See, for example, "O zemnykh aliumofosfornykh i aliumosernykh analogakh kaolinovykh aliumosilikatov," *Doklady Akademii nauk,* v. 18, nos. 4–5(1938): 621ff.

84. A. A. Kornilov, *Sem' mesiatsev sredi golodaiushchikh krest'ian,* Moscow, 1893, pp. 8–9.

85. Kornilov, 1893, p. 227–29.

86. See Richard G. Robbins, Jr. *Famine in Russia 1891–1892: The Imperial Government Responds to a Crisis,* New York, 1975.

87. Kornilov, pp. 12–14.

88. Kornilov, 1893, pp. 85–89.

89. A. A. Kornilov, "Vospominaniia," AAN f. 518, op. 5, d. 68, 1. 201.

90. Kornilov, 1893, pp. 140–42.

91. For Vernadsky's jaundiced view of the village clergy, see his letter to his wife, 4 June 1886. Arkhiv Akademii Nauk SSSR.

92. Kornilov, 1893, pp. 215–22.

93. The most detailed recent account of that role can be found in Terence Emmons' splendid monograph, *The Formation of Political Parties and the First National Elections in Russia,* Harvard, 1983, especially pp. 21–88 and 145–205.

94. Mochalov, 1982, p. 111.

95. V. I. Vernadsky to N. E. Vernadskaia, 25 October 1892.

96. Mochalov, 1982, p. 109.

97. See P. I. Shlemin. "Zemsko-liberal'noe dvizhenie i adresa 1894/95 g.," *Vestnik Moskovskogo universiteta: Istoriia,* 1 (1973): 197–99. I have been unable to find any other mention of this in primary or secondary sources.

98. V. I. Vernadsky to N. E. Vernadskaia, 24 March 1892.

99. V. I. Vernadsky to N. E. Vernadskaia, 13 April 1892.

100. Mochalov, 1982, pp. 116–19.

101. Emmons, 1983, p. 64-65.

102. George Vernadsky, "Notes and Corrections on the Book of Lev Gumilevskii," BAR, Vernadsky Collection, Box 29.3.5.2., Folder 115, Columbia University.

103. Portions of the police dossier on Vernadsky for this period were published in *Voprosy istorii estestvoznanii i tekhniki,* Moscow, 15(1963): 122–23.

104. Gumilevskii, p. 73.

105. See, for example, Dokuchaev to Vernadsky, 11 April 1892; see also Vernadsky to Dokuchaev, 27 January 1892, in Dokuchaev correspondence, pp. 827–28.

106. V. I. Vernadsky to N. E. Vernadskaia, 8 March and 21 March 1892.

107. See Groth's letter to Vernadsky of 5 April 1892, Kabinet-Muzei V. I. Vernadskogo, Akademiia Nauk SSSR, Moscow. Groth ceased correspondence with Vernadsky until they reestablished friendly ties during Groth's visit in 1897 to attend a geological congress in Moscow.

108. N. E. Vernadskaia to A. V. Gol'shtein, 18 January 1892, BAR, Vernadsky Collection, Columbia University.

109. V. I. Vernadsky to N. E. Vernadskaia, 3 May 1892; Mochalov, 1982, pp. 103–4.

110. *Dictionary of Scientific Biography,* vol. 13, p. 617.

111. V. I. Vernadsky to N. E. Vernadskaia, 29 June 1896, Arkhiv AN SSSR.

112. V. I. Vernadsky to N. E. Vernadskaia, 7 July 1896, BAR, Vernadsky Collection, Columbia University.

113. V. I. Vernadsky to N. E. Vernadskaia, 4 September 1896, BAR, Columbia University.

114. V. I. Vernadsky to N. E. Vernadskaia, 4 September 1896, BAR, Columbia University.

115. V. I. Vernadsky to N. E. Vernadskaia, 7 July 1896, BAR, Vernadsky Collection, Columbia University.

116. Ibid.

117. V. I. Vernadsky to N. E. Vernadskaia, 2 September 1896, BAR, Columbia University.

118. V. I. Vernadsky to N. E. Vernadskaia, 8 July 1896, BAR, Columbia University.

119. V. I. Vernadsky to N. E. Vernadskaia, 14 October 1896, BAR, Columbia University.

120. V. I. Vernadsky to N. E. Vernadskaia, 2 July 1896, BAR, Columbia University.

121. I. I. Shafranovskii, "Raboty V. I. Vernadskogo po kristallografii," *Zapiski vsesoiuznogo mineralogicheskogo obshchestva,* vypusk 1, chast' 75, Moscow, 1946.

122. N. M. Raskin and I. I. Shafranovskii, "E. S. Fedorov i V. I. Vernadskii (Po materialam arkhiva Akademii nauk SSSR)," *Ocherki po istorii geologicheskikh znanii,* Moscow, 8 (1959): 166.

123. Raskin and Shafranovskii, op. cit., p. 168, citing a reveiw of 1898 by Fedorov in *Ezhegodnik po geologii i mineralogii Rossii,* pp. 11-18.

124. A. E. Fersman, "Vladimir Ivanovich Vernadskii," *Biulleten' Moskovskogo obshchestva ispytatelei prirody, otdeleni geologii,* 21 (1946): 54.

125. V. I. Vernadsky, *Osnovy kristallogradii,* Moscow, 1903.

126. Mochalov, 1982, pp. 142–44.

127. Balandin, p. 41.

128. Balandin, p. 43.

129. Balandin, p. 44.

130. Balandin, p. 49.

131. Shafronovskii, op. cit.

132. Mochalov, 1982, p. 138.

133. Raskin and Shafranovskii, op. cit., 1959, pp. 172–73.

134. Six volumes of this work were published in small editions by the Academy of Sciences between 1908 and 1922: See *Opyt opisatel'noi mineralogii,* St. Petersburg, 1908, 1909, 1910, 1912; Petrograd, 1914, 1918, 1922.

135. I. I. Shafranovskii, *Evgraf Stepanovich Fedorov,* Moscow, 1963, pp. 1–10.

136. *Dictionary of Scientific Biography,* vol. 5, New York, 1972, pp. 210–14.

137. Raskin and Shafranovskii, p. 173.

138. Vernadsky to N. E. Vernadskaia, April 1903; cited in Mochalov.

139. Ibid.

140. E. E. Flint, "Vospominaniia."

141. A. E. Fersman, *Moi puteshestviia,* Moscow, 1949.

142. Ibid.

143. Ibid.

144. A. E. Fersman, "Vladimir Ivanovich Vernadskii," *Biulleten' Moskovskogo obshchestva ispytatelei prirody, otdeleni geologii,* 21 (1946): 54.

145. P. K. Kazakova, "Vospominaniia," Kabinet-muzei Vernadskogo, Moscow.

146. Mochalov, 1982, pp. 142–44.
147. Mochalov, 1982, p. 130.
148. E. D. Revutskaia, "Vospominaniia," Kabinet-muzei Vernadskogo, Moscow.
149. O. M. Shubnikova, "Akademik V. I. Vernadskii i Professor Iakov V. Samoilov," *Ocherki po istorii geologicheskikh znanii,* Moscow, 1953.
150. *Dictionary of Scientifc Biography.*
151. Shubnikov, op. cit.
152. Mochalov, 1982, p. 120.
153. Mochalov, 1982, p. 120.
154. Mochalov, 1982, p. 127.
155. Vernadsky to N. E. Vernadskaia, 1903; cited in Mochalov, p. 139.
156. O. N. Pisarzhevskii, *Fersman,* Moscow, 1959.

3. THE POLITICS OF MORAL INDIGNATION

1. V. I. Vernadsky to V. A. and A. V. Gol'shtein, Vernadsky Collection, BAR, Columbia University.
2. See, for example, Vernadsky's letters to his wife, Natasha E. Vernadskaia, dated 7 May 1901 and 6 September 1901, Kabinet-muzei Vernadskogo, Institut geokhimii, Moscow.
3. V. I. Vernadsky, *O nauchnom mirovozzrenii,* first published in *Voprosy filosofii i psikhologii* 65 (1 August 1902), Kabinet-muzei Vernadskogo, Moscow.
4. V. I. Vernadsky, *Ocherki i rechi,* Vol. II, Petrograd, 1922, pp. 21, 26, 30.
5. V. I. Vernadsky, "Tri resheniia," *Poliarnaia zvezda* 14 (1906): 172.
6. "Dnevnik, 1890–1894," AAN f. 518. op. 2, d. 5, 11. 49–50; Mochalov, p. 114.
7. "Iz zapisok 1892," AAN f. 518, op. 1, d. 215, 11. 2, 3, cited in Mochalov, p. 117.
8. V. I. Vernadsky to N. E. Vernadskaia, 1 July 1893, AAN f. 518, op. 7, d. 40, 11. 51–52, cited in Mochalov, p. 118.
9. Mochalov, p. 117.
10. V. I. Vernadsky to N. E. Vernadskaia, 10 June 1894, AAN f. 518, op. 7, d. 40, 1. 29, cited in Mochalov, p. 118.
11. V. I. Vernadsky to N. E. Vernadskaia, 10 June 1894, AAN f. 518, op. 7, d. 41, 1. 18, cited in Mochalov, p. 118.
12. Richard Pipes, *Struve: Liberal on the Left 1870–1905* Cambridge, Mass., 1970, pp. 336–37.
13. "Progress nauki i narodnye massy," in V. I. Vernadsky, *Izbrannye trudy po istorii nauki,* Moscow, 1981, pp. 186–87.
14. Pipes, *Struve,* pp. 322, 327.
15. V. I. Vernadsky to N. E. Vernadskaia, 18 July 1903, AAN f. 518, op. 7, d. 49, 1. 39, cited in Mochalov, p. 140–41.
16. Emmons, pp. 3, 68.
17. Manning, p. 61.
18. N. A. Balashova, *Rossiiskii liberalizm nachala XX-ogo veka,* Moscow, 1981, p. 52.
19. *Sbornik statisticheskikh svedenii po Tambovskoi gubernii,* Vol. 15, *Chastnoe zemlevladenie Morshanskogo uezda,* published by the Tambov provincial zemstvo, Tambov, 1890, pp. 130–42, cited in I. I. Mochalov, "Estestvennonauchnye i filosofskie osnovy mirovozzrenie V. I. Vernadskogo," doctoral dissertation in Lenin Library, Moscow, 1969, p. 315.
20. Mochalov dissertation, p. 315.
21. V. I. Vernadsky to N. E. Vernadskaia, 11 July 1896.
22. Ibid.

23. Mochalov (1982), pp. 109–111; G. Vernadsky, *Novye zhurnal*.

24. Ibid.

25. V. I. Vernadsky to N. E. Vernadskaia, 25 October, 1892.

26. Mochalov (1982), p. 109.

27. V. I. Vernadsky to N. E. Vernadskaia, 26 September 1900.

28. V. I. Vernadsky to N. E. Vernadskaia, December 1892, Kabinet-muzei Vernadskogo, Moscow.

29. V. I. Vernadsky to M. A. D'iakonov, 27 December 1892, AAN f. 69, d. 125, 1. 2, cited in Mochalov (1982), p. 111.

30. V. I. Vernadsky to N. E. Vernadskaia, 26 September 1900 and 3 December 1900, Kabinet-muzei Vernadskogo, Moscow.

31. V. I. Vernadsky to N. E. Vernadskaia, 5 August 1900, Kabinet-muzei Vernadskogo, Moscow.

32. Vernadsky's diary, 1899, AAN f. 518, op. 1, d. 4, 1. 110, cited in Mochalov (1982), p. 121.

33. V. I. Vernadsky to N. E. Vernadskaia, 26–29 September 1900, AAN f. 518, op. 7

34. V. I. Vernadsky to N. E. Vernadskaia, October 1899, AAN f. 518, op. 2, khr. 1899, p. 1; letter to N. E. Vernadskaia, 3–6 December 1904, AAN f. 518, op. 2, khr. 1904, p. 2.

35. Emmons, p. 69.

36. Manning, p. 47.

37. Manning, p. 60.

38. Manning, p. 12.

39. Manning, p. 20.

40. Emmons, p. 30.

41. Manning, p. 88.

42. Manning, pp. 71–77.

43. Emmons, p. 33.

44. See, for example, Vernadsky's letter of September 30, 1905, to his wife, written from Morshansk county where he was already reporting that the mood of the local gentry had moved to the right. Vernadsky felt he might not be reelected to the zemstvo, since there was agitation against him and one other local candidate. Vernadsky indicated that he favored concessions to the peasants but was not optimistic about the chances of their succeeding, given the rising tide of reaction among the local nobility.

45. Manning, pp. 125–32, 187–95.

46. Kassow, p. 156.

47. AAN, f. 518, op. 2, d. 4, 11. 105, 108, 111; cited in Mochalov (1982), p. 123.

48. Kassow, pp. 169–71.

49. Kassow, p. 179.

50. Kassow, p. 179.

51. TsGIA, f. 7, op. 1515, d. 117, 1. 41 (Ministerstvo vnutrennikh del. Kharakteristika na V. I. Vernadskogo, 1899) and TsGAM, f. 459, op. 2, d. 4802 (Kantseliariia popechitelia Moskovskogo uchebnogo okruga); cited in Mochalov (1982), pp. 123–24.

52. Mochalov (1982), p. 124.

53. V. I. Vernadsky, *Ob osnovaniiakh universitetskoi reformy*, Moscow, 1901.

54. V. I. Vernadsky to N. E. Vernadskaia, 5 May 1901.

55. V. I. Vernadsky to N. E. Vernadskaia, 8 May 1901.

56. V. I. Vernadsky to N. E. Vernadskaia, 1 May 1901.

57. V. I. Vernadsky to N. E. Vernadskaia, 26 August 1901 and 6 September 1901.

58. V. I. Vernadsky to N. E. Vernadskaia, 29 April 1902; McClelland, p. 40.

59. Kassow, p. 197.

60. Kassow, pp. 198–205.

61. Kassow, p. 208.

62. Kassow, pp. 209–16.
63. Kornilov, p. 8.
64. Galai, p. 199.
65. Kassow, p. 300–301.
66. Kassow, pp. 301–2, citing a letter in Ol'ga Trubetskaia, *Kniaz' S. N. Trubetskoi: Vospominaniia sestry*, New York, 1953, p. 74.
67. V. G. Glazov, "Dva razgovora: iz dnevikov V. G. Glazova," *Dela i dni*, 1 (1920): 209; cited in Kassow, pp. 220–21.
68. Kassow, pp. 220–21.
69. Kassow, pp. 225–26.
70. Kassow, p. 295.
71. Manning, p. 68.
72. Shipov, p. 281. On Kryzhanovsky's role as a member of the Oldenburg group in the 1880s, see Chapter One.
73. Emmons, p. 9.
74. G. Vernadsky, "Bratstvo 'Priutino,'" p. 169; Galai, p. 223; Kassow, p. 309; Emmons, p. 32.
75. Emmons, p. 165.
76. Emmons, p. 9.
77. Mochalov (1982), p. 148; Kassow, pp. 286, 304–5.
78. *Nashi dni*, 18 December 1904.
79. See, for example, the letter to Vernadsky from the young Moscow geneticist, N. K. Kol'tsov, in which he reports widespread support among faculty for the idea of a national congress of professors: N. K. Kol'tsov to V. I. Vernadsky, 28 December 1904, first published in *Genetika* vol. IV (no. 4, 1968): 148–49.
80. Kassow, p. 258.
81. Kassow, pp. 259–60.
82. Kassow, p. 282.
83. *Russkie vedomosti*, 29 January 1905.
84. Sidney Harcave, *The Russian Revolution of 1905*, p. 119.
85. Kassow, pp. 277–283.
86. Kassow, p. 323.
87. Kassow, p. 323.
88. Letter of N. K. Kol'tsov to V. I. Vernadsky, 13/26 May 1905, first published in *Genetika* vol. IV (no. 4, 1968): 149–50.
89. *Pravitel'stvennyi vestnik*, 23 April 1905.
90. N. K. Kol'tsov to V. I. Vernadsky, 13/26 May 1905, pp. 149–50.
91. See, for example, the article in the German periodical *Akademische Revue*, June 1905, p. 275.
92. V. I. Vernadsky, "Blizhaishie zadachi akademicheskoi zhizni," *Pravo*, 19 June 1905 and "Tri zabastovki," *Russkie vedomosti*, 9 August 1905.
93. V. I. Vernadsky, "Tri zabastovki," *Russkie vedomosti*, 9 August 1905.
94. "Prilozhenie," *Osvobozhdenie*, nos. 78/79 (1905): 14.
95. Kassow.
96. *Russkie vedomosti*, 29 August 1905, summarized in Kassow, pp. 342–43.
97. See N. Speranskii, *Vozniknovenie Moskovskogo gorodskogo narodnogo universiteta imeni A. L. Shaniavskogo*, Moscow, 1913.
98. O. Pisarzhevskii, *Fersman*, Moscow, 1959, pp. 129–31.
99. V. I. Vernadsky to N. E. Vernadskaia, 4 July 1905.
100. N. K. Kol'tsov, *K universitetskomu voprosu*, Moscow, 1909.
101. Kassow, p. 288.
102. Kassow, pp. 345–46.
103. Kassow, p. 345.

104. Mochalov (1982), p. 151.

105. V. I. Vernadsky to N. E. Vernadskaia, 4 July 1905, 19 August 1905; Paul Miliukov, *Political Memoirs, 1905–1917*, Ann Arbor, Michigan, 1967, pp. 42–43.

106. Kassow, p. 338.

107. Kassow, p. 407.

108. Kassow, p. 410.

109. Kassow, p. 411, relying on Latysheva, p. 242.

110. A. A. Kornilov, "Vospominaniia," unpublished manuscript, Kabinet muzei Vernadskogo, p. 411.

111. Kornilov, p. 421.

112. Kassow, p. 417.

113. Manning, p. 140.

114. Kassow, p. 424; Kornilov autobiography, unpublished manuscript, Kabinet-muzei, pp. 425–26.

115. *Russkie vedomosti*, 8 November 1905, cited in Kassow, p. 429.

116. Kassow, pp. 418–24.

117. Kornilov, "Vospominaniia," p. 423.

118. G. Vernadsky, "Bratstvo 'Priutino'," *Novyi zhurnal*, kn. 97, New York, 1969, p. 219.

119. V. I. Vernadsky to N. E. Vernadskaia, 2 February 1906, Kabinet-muzei.

120. Kornilov, "Vospominaniia," p. 421.

121. "Prilozhenie," *Osvobozhdenie*, 78/79 (1905): 1–14.

122. Emmons, p. 147.

123. Emmons, p. 193.

124. See the brochure Vernadsky published on this topic in 1905: *Zapiska ob otnoshenii Moskovskogo universiteta k "Moskovskim vedomostiam" i k universitetskoi tipografii*, Moscow, 1905.

125. See, for example, Mochalov (1982), pp. 152–53 and Gumilevskii, pp. 73–80, who ignores Vernadsky's important role in the Kadet party altogether.

126. See the letter signed by George Vernadsky, 'student at Moscow University', in *Russkie vedomosti*, 18 January 1906.

127. G. Vernadsky autobiography, unpublished, BAR, Columbia University.

128. V. I. Vernadsky to N. E. Vernadskaia, 2 February 1906.

129. V. I. Vernadsky to N. E. Vernadskaia, 2 February 1906; George Vernadsky autobiography, op. cit.

130. Emmons, p. 182.

131. Emmons, pp. 187–92.

132. Emmons, p. 185.

133. Emmons, p. 160; Manning, pp. 201–6.

134. V. I. Vernadsky to N. E. Vernadskaia, 19 June 1906.

135. Manning, p. 276.

136. *Gosudarstvennyi sovet. Stenograficheskii otchet. Sessiia 1-ia, zasedanie 11*, St. Petersburg, 1906, pp. 30–31.

137. "Zapiski 1905, 1942," AAN, f. 518, op. 2, d. 4, 1. 136; Mochalov, p. 139.

138. Mochalov, pp. 251–52.

139. "Dnevnik, 1906," AAN, f. 518, op. 2, d. 4, 11. 168–69.

140. Manning, pp. 175–76.

141. Ibid.

142. Manning, p. 173.

143. V. I. Vernadsky to G. V. Vernadsky, 14 March 1906, TsGAOR, f. 117, op. 1, d. 200, 1. 25.

144. Gumilevskii, p. 78.

145. Mochalov (1982), p. 185.

146. V. I. Vernadsky to Ia. V. Samoilov, 1 August 1906, AAN, f. 518, op. 2, d. 1997, 1. 25.

147. V. I. Vernadsky to G. V. Vernadsky, 20 May 1906, TsGAOR, f. 117, op. 1, d. 200, 1. 26.

148. "Pisma o Gosudarstvennom sovete," *Russkie vedomosti,* 24 May 1906.

149. Ibid.

150. V. I. Vernadsky to N. E. Vernadskaia, 28 June 1906, AAN, f. 518, op. 2, d. 7, 1. 82.

151. V. I. Vernadsky to N. E. Vernadskaia, 2 July 1906.

152. *Rech',* 10 July 1906.

153. Kornilov autobiography, p. 448.

154. V. I. Vernadsky to N. E. Vernadskaia, 1 July 1906, AAN, f. 518, op. 2, d. 7, 1. 86. Some sources, like Gumilevskii, contend that Vernadsky signed the appeal, but I have been unable to find his name among the signers. In a handwritten statement commenting on errors in Gumilevskii's book, George Vernadsky states that his father, who was not a member of the Duma, did not sign the appeal. See George Vernadsky's comments on the Gumilevskii biography, BAR, Vernadsky Collection, Columbia University.

155. Kassow, p. 447.

156. Vernadsky was not a delegate to the Tolstoy conference, and I have been unable to ascertain his position on this question. However, his close friend, Professor Ivan Grevs of St. Petersburg University, led a bloc at this conference which favored giving junior faculty voting rights in the councils, but not a decisive vote. Their position was resoundingly defeated, however. Given Vernadsky's general worldview and position on other questions, it is reasonable to assume that he shared the views of Grevs and other liberal professors at this conference. See also, Kassow, pp. 460–61.

157. McClelland, p. 66; Kassow, pp. 467–68.

158. "Zapiski, 1907," cited in Mochalov (manuscript), p. 268.

159. McClelland, p. 66.

160. Kassow, pp. 474, 606.

161. See, for example, Vernadsky's article in *Ezhegodnik gazety Rech',* St. Petersburg, 1914.

162. Kassow, p. 542.

163. *Rech',* 12 January 1908.

164. Kassow, p. 544.

165. Kassow, p. 538.

166. Kassow, p. 597.

167. Kassow, p. 548–60.

168. Kassow, p. 642.

169. *Russkie vedomosti,* 1 February 1911.

170. *Russkaia mysl',* 3 (1911): 162, cited in Kassow, p. 660.

171. Kassow.

172. McClelland, pp. 41, 87, 105.

173. *Vestnik vospitaniia,* 2 (1911), as cited in Kassow, p. 676.

174. V. I. Vernadsky, "K voprosu o rasprostranenii skandiia," *Izvestiia AN,* vol 2 (no. 17, 1908):1273–74.

175. V. I. Vernadsky, "O tsezii v polevykh shpatakh," *Izvestiia AN,* vol. 3 (no. 3, 1909):163–64.

176. V. I. Vernadsky, "Zametki o rasprostranenii khimicheskikh elementov v zemnoi kore," *Izvestiia AN,* 6-ia seriia, vol. 8 (no. 1, 1914), republished in V. I. Vernadsky, *Izbrannye sochinenii,* Moscow, 1954, pp. 460–71.

177. R. Balandin, 1979, p. 57.

178. V. I. Vernadsky, "Paragenezis khimicheskikh elementov v zemnoi kore," *Dnevnik XII s"ezda russkikh estestvoispytateli i vrachei,* Moscow, 1910, pp. 73–91, reprinted in Vernadsky, *Izbrannye sochinenii,* Moscow, 1954, pp. 404–7.

179. V. M. Korsynskaia and N. M. Berzilin, *V. I. Vernadskii,* Moscow, 1975, pp. 61–64.

180. Korsynskaia and Berzilin, p. 404.

181. Korsynskaia and Berzilin, p. 407.

182. V. I. Vernadsky to N. E. Vernadskaia, 6 September 1911, Kabinet muzei Vernadskogo.

183. V. I. Vernadsky, "O gazovom obmene zemnoi kory," *Izvestiia AN,* 1912, republished in *Izbrannye sochinenii,* pp. 90–112.

184. V. I. Vernadsky to Ia. V. Samoilov, July 1909.

185. "Mysli, 1901–1911," AAN, F. 518, op. 1, d. 161, 1. 101.

186. Ibid.

187. V. I. Vernadsky to G. V. Vernadsky, 27 June 1908, TsGAOR, f. 117, op. 1, d. 200, 1. 48, as cited in Mochalov, p. 172.

188. Balandin, pp. 54–56.

189. V. I. Vernadsky to N. E. Vernadskaia, 11 September 1911, Kabinet-muzei Vernadskogo.

190. Vernadsky's views were first presented at a student memorial meeting for Trubetskoi on 16 March 1908, and were published in 1922 as an article in an anthology of Vernadsky's work: "Cherty mirovozzreniia Kniaz'ia S. N. Trubetskogo," *Ocherki i rechi Akademika V. I. Vernadskogo,* Vol. II, Petrograd, 1922, pp. 95–96.

191. V. I. Vernadsky to Ia. V. Samoilov, 9 July 1908, quoted in Shakhovskaia "khronologiia," Kabinet muzei Vernadskogo, Moscow.

192. V. I. Vernadsky to N. E. Vernadskaia, 28 December 1910, Kabinet-muzei Vernadskogo.

193. V. I. Vernadsky, "Zametki o rasprostranenii khimicheskikh elementov v zemnoi kore," p. 462.

194. V. I. Vernadsky, "O neobkhodimosti issledovaniia radioaktivnykh mineral Rossiiskoi Imperii," St. Petersburg, 1914, reprinted in V. I. Vernadsky, *Izbrannye sochinenii,* pp. 574–80.

195. V. I. Vernadsky to N. E. Vernadskaia, 21 May 1911, Kabinet-muzei Vernadskogo.

196. V. I. Vernadsky to N. E. Vernadskaia, 4 May 1911, TsGAOR, f. 1137, op. 1, d. 485, 1. 39.

197. V. I. Vernadsky to N. E. Vernadskaia, 18 May 1911, Kabinet-muzei Vernadskogo.

198. Yegorov, 1914, p. 60.

199. Mochalov, p. 325; Yegorov, p. 60.

200. V. I. Vernadsky to A. E. Fersman, 4 June 1911, published in *Aleksandr Evgenevich Fersman. Zhizn' i deiatel'nost,* Moscow, 1965, p. 323.

201. See, I. V. Paramonov and N. P. Korobochkin, *Nikolai Mikhailovich Fedorovskii, 1886–1956,* Moscow, 1979, pp. 20–21.

202. See George Greenstein, "Heavenly Fire," *Science* (July/August 1985): 70–77.

203. Paramonov and Korobochkin, pp. 8–23.

204. A. B. Missuna to V. I. Vernadsky, 8 March and 19 December 1912, AAN, f. 518, op. 2, no. 1080, summarized in Shakhovskaia khronologiia, 1912, Kabinet-muzei. Out of gratitude, Missuna later named a new mineral she had discovered in 1914 after Vernadsky, see AAN, f. 518, op. 2, khr. 194, p. 13.

205. V. I. Vernadsky to N. E. Vernadskaia, 2 August 1913.

206. V. I. Vernadsky to N. E. Vernadskaia, 24 August 1913; V. I. Vernadsky to N. E. Vernadskaia, 5 May 1913.

207. Ibid.

208. V. I. Vernadsky to N. E. Vernadskaia, 27 October 1913.

209. George Vernadsky autobiography, BAR, Columbia University.

210. V. I. Vernadsky to N. E. Vernadskaia, May 1913.

211. L. V. Vasil'eva, "Vospominaniia," unpublished manuscript in the Kabinet-muzei Vernadskogo, Moscow.

212. Vasil'eva, "Vospominaniia."

213. A. S. Krylenko to V. I. Vernadsky, 6 August 1911, Kabinet-muzei Vernadskogo, Moscow.

214. P. K. Kazakova, "Vospominaniia," unpublished manuscript, Kabinet-muzei Vernadskogo, Moscow.

215. E. E. Flint, "Vospominaniia," unpublished manuscript, Kabinet-muzei Vernadskogo, Moscow.

4. SCIENCE, WAR, AND REVOLUTION

1. Mochalov, 1982, p. 305.

2. Gumilevskii, p. 119.

3. A. A. Kornilov, "Vospominaniia," manuscript. AAN f. 518, op. 5, d. 68, 1915, 1916.

4. "Voina i progress nauki," Petrograd, 1915, reprinted in *Ocherki i rechi Akad. V. I. Vernadskogo,* Vol. 1, Petrograd, 1922, pp. 128–40.

5. Vernadsky to Fersman, August 1915, published in *Pis'ma V. I. Vernadskogo A. E. Fersmanu,* Moscow, 1985, p. 7.

6. Perel'men, 1983, p. 35.

7. A. N. Krylov, *Vospominaniia,* Moscow, 1942.

8. Mochalov, 1982, p. 207.

9. AAN, f. 518, op. 2, khr. 1915, p. 3.

10. A. E. Fersman, "V. I. Lenin i izuchenie proizvoditel'nykh sil SSSR," in the book *Problemy mineral'nogo syr'ia,* Moscow, 1975, p. 4.

11. V. I. Vernadsky to A. V. Gol'shtein, 26 November 1916, BAR, Columbia University.

12. V. I. Vernadsky to George Vernadsky, 25 July 1917, TGAOR, f. 1137, op. 1, ed. khr. 200.

13. V. I. Vernadsky to Ia. V. Samoilov, 8 July 1917, AAN. f. 518, op. 2, d. 2002, 1. 7, cited in Mochalov, 1982, p. 218.

14. AAN, f. 518, op. 2, d. 4, 1. 220, Dnevnik, 1917, cited in Mochalov, 1982, p. 220.

15. Gumilevskii, p. 127.

16. Kabinet-muzei, V. I. Vernadsky, *Iz vospominanii,* 1943.

17. Praskovia Kirillovna Kazakova, "Vospominaniia," Kabinet muzei, Moscow.

18. V. I. Vernadsky to Ia. Samoilov, 2 June 1918, quoted in Shakovskaia, "Khronologiia," 1918, Kabinet-muzei, Moscow.

19. Mochalov, 1982, pp. 213–25.

20. Sergei Timoshenko, *As I Remember,* Princeton, 1968, pp. 166ff.

21. Vernadsky, "Vospominaniia," 1943, Kabinet-muzei, Moscow.

22. AAN, f. 518, op. 2, ed. khr. 11.

23. Diary entry of 19 April 1919, AAN, f. 518, op. 2, ed. khr. 11.

24. Diary entry of 1 November 1919, AAN, f. 518, op. 2, ed. khr. 11.

25. Diary entry of 14 October 1919, AAN, f. 518, op. 2, ed. khr. 11.

26. Diary entry of 24 November 1920, Kabinet-muzei, Moscow.

27. Diary entry of 29 December 1919, AAN, f. 518, op. 2, ed. khr. 11.

28. Diary entries of 21 August 1920 and 9 September 1920.

29. As he put it in a 1923 letter to Fedor Rodichev, "I myself am not Orthodox, not even a Christian, but I consider myself closer, immeasurably closer, to any Orthodox believer than to a believing socialist. The latter are in my opinion dangerous for freedom and the development of humanity." V. I. Vernadsky to F. I. Rodichev, 10 March 1923, BAR, Columbia University.

30. Letter of Nina Vernadskaia Toll to N. S. Neapolitanskaia, 7 September 1975, Kabinet-muzei, Moscow.

31. V. I. Vernadsky, Diary entry of 24 November 1920, AAN, f. 518, op. 2, ed. khr. 11.

32. Theodosius Dobzhansky, "Memoirs," Oral History Collection, Columbia University Library, vol. 1, pp. 57–71.

33. N. G. Kholodny, "Iz vospominanii o V. I. Vernadskom, *Pochvovedenie* 7(1945); *Mysli naturalista o prirode i cheloveke* Kiev, 1947; *Izbrannye trudy*, 3 vols., Kiev, 1956–57; *Zhelezobakterii*, Moscow, 1953. Besides Kholodny, there were also A. V. Fomin, M. I. Bessmertnoi and I. V. Starynkevich, who considered themselves part of Vernadsky's school. See Mochalov, 1982, p. 225; Gumilevskii, p. 123.

34. Praskovia Kirillovna Kazakov, "Vospominaniia," manuscript, Kabinet-muzei, Moscow.

35. Nina Vernadskaia, "Memoirs," Hoover Institution Archives, Stanford, p. 27.

36. Diary entry 27 November 1920.

37. *Perepiska V. I. Vernadskogo s B. L. Lichkovym 1918–1939*, Moscow, 1979, p. 8; *Pis'ma V. I. Vernaskogo A. E. Fersmanu*, Moscow, 1985, p. 100.

38. George Vernadsky, *Memoirs*, Bakmeteff Archive, Columbia University.

39. V. I. Vernadsky, diary entry 1920, AAN, f. 518; Mochalov, 1982, pp. 228–32.

40. E. E. Flint, "Vospominaniia," manuscript, Kabinet-muzei, Moscow.

41. Unpublished letter in TsGAOR, f. 2306, op. 2, ed. khr. 784, l. 114.

42. V. I. Vernadsky to B. L. Lichkov, 28 April 1921, in *Perepiska V. I. Vernadskogo s B. L. Lichkovym, 1918–1939*, Moscow, 1979.

43. *Otchet o deiatel'nost Rossiskoi Akademii nauk*, Petrograd, 1917, pp. 1–5.

44. "Vospominaniia akademika S. F. Oldenburga o vstrechakh s V. I. Leninym v 1887 i 1921 godakh." in *Lenin i Akademiia nauk*, pp. 88–94. The great impression Lenin made on Oldenburg is corroborated by the memoirs of his friend, V. I. Vernadsky, and by Lenin's secretary, V. D. Bonch-Bruevich. Vernadsky's comment is to be found in AAN, f. 518, op. 4, ed. khr. 45, l. 12 and Bonch-Bruevich's in *Lenin i Akademiia nauk*, pp. 25–26.

45. Kol'tsov, p. 34.

46. Seven were in the physical and mathematical sciences, four in chemistry, four in the earth sciences, and five in biological sciences. AAN f. 410, op. 1, ed. khr. no. 94, 1. 4, cited in N. M. Mistriakva, "Struktura, nauchnye uchrezhdeniia i kadry AN SSSR (1917–1940 gg.)," *Organizatsiia nauchnoi deiatel'nosti*, Moscow, 1968, p. 214.

47. L. V. Ivanova, *Formirovanie sovetskoi nauchnoi intelligentsii 1917–1927*, Moscow, 1950, p. 39.

48. *Vestnik Akademii nauk*, 8(1967): 69–70.

49. The report of this meeting is contained in TsGAOR RSFSR, f. 2306, op. 19, ed. khr. 3, l. 86, cited in Ivanova, p. 37.

50. Kol'tsov, p. 41.

51. George Leggett, *The Cheka: Lenin's Political Police*, Oxford, 1981, pp. 306–308.

52. See the articles by Lunacharsky in *Novosti dni* 5 April 1918, p. 1 and a more detailed account of Narkompros' negotiations with the Academy in *Izvestiia* 12 April 1918, p. 3. The Sovnarkom decree, signed 12 April 1918, was first published in *Izvestiia* 19 April 1918.

53. *Ekonomicheskaia gazeta* 21 July 1931, p. 3.

54. V. A. Ulianovskaia, *Formirovanie nauchnoi intelligentsia v SSSR 1917–1937*, Moscow, 1966, p. 59.

55. *Narodnoe khoziastvo* 1(1918):3–6.

56. See, for example, Alexander Fersman in *Nauka i ee rabotniki* 1(1921): 3ff.

57. Ibid.

58. Kol'tsov, pp. 99, 111, 123–24, 133.

59. "Nabrosok plana nauchno-tekhnicheskikh rabot," April 18–25, 1918, reprinted in *Lenin i Akademiia nauk*, pp. 44–47.

60. See the documents in *Organizatsiia nauki v pervye gody*, Moscow, pp. 113–23.

61. See Sheila Fitzpatrick, p. 75.

62. Bastrakova, p. 212, citing TsGAOR RSFSR, f. 2306, op. 1, ed. khr. 35, 11. 95–96.

63. AAN f. 162, op. 3, ed. khr. 171, 11. 41, 51–52 ob., st. 56; also cited in Bastrakova, p. 100.

64. TsGAOR RSFSR, f. 2306, op. 19, ed. khr., 18, 1. 198, cited in Bastrakova, pp. 100–101.

65. Ibid.

66. TsGANKh SSSR, f. 3429, op. 60, ed. khr. 30, 11. 207, 207 ob., cited in Bastrakova, p. 213.

67. See, for example, Fitzpatrick, pp. 71–72.

68. This letter is published in *Lenin i Akademiia nauk. Sbornik dokumentov*, Moscow, p. 61.

69. *Novyi mir* 10 (1925): 110; Kol'tsov, p. 63.

70. On the Proletkult during this period see, for example, V. V. Gorbunov, "Iz istorii bor'by Kommunisticheskoi partii s sektanstvom Proletkul'ta," in *Ocherki po istorii sovetskoi nauki i kul'tury*, Moscow, 1968, pp. 29–68; and Fitzpatrick, pp. 89–109, 178–80, 185–87, 238–41, 169–70.

71. Kol'tsov, p. 177.

72. Ivanova, p. 360; Bastrakova, p. 162.

73. Fersman, p. 6.

74. See Kol'tsov, pp. 99, 111, 123–24, 133.

75. Bonch-Bruevich, op. cit., pp. 68–69.

76. *Pis'ma V. I. Vernadskogo A. E. Fersmanu*, Moscow, 1985, p. 100; Mochalov, 1982, p. 233.

77. V. I. Vernadsky to Ivan Il'ich Petrunkevich, 10 March 1923, Vernadsky Collection, BAR, Columbia University.

78. Vernadskii manuscript autobiography, BAR, Columbia University, Box 31.8.6.I-I-6, Folder I-3-A-a-1-1.

5. THE VERNADSKY SCHOOL AND SOVIET SCIENCE

1. A. V. Gol'shtein to G. V. Vernadsky, 21 October 1921, Gol'shtein collection, BAR, Columbia University.

2. A. V. Gol'shtein to G. V. Vernadsky, 29 October 1925; Gol'shtein to N. E. Vernadskaia, 12 December 1925, BAR, Columbia University.

3. V. I. Vernadsky to F. I. Rodichev, 20 September 1929, BAR, Columbia University.

4. V. I. Vernadsky to A. E. Fersman, 3 August 1927, in *Pis'ma V. I. Vernadskogo A. E. Fersmanu*, pp. 116–17.

5. See, for example, Vernadsky Papers, BAR, Box 25.5.9.1, Folder 15; Box 25.5.9.1, Folder 12; Box 25.4.7.2, Folder 15.

6. Vernadskii manuscript autobiography, Bakhmeteff Archive.

7. Ibid.

8. V. I. Vernadsky to Fedor Rodichev, undated but written toward the end of 1925, BAR, Box 3012.3.2.4, Folders 97–98.

9. See the letters to Fersman, 20 June 1925; 10 September 1925; 5 January 1926; 7 January 1926; 4 February 1926, in *Pis'ma V. I. Vernadskogo A. E. Fersmanu*, pp. 119–31.

10. Vernadsky to Fersman, 2 November 1923, in *Pis'ma*, pp. 111–12.

11. See A. P. Vinogradov, *Chelovek, obshchestvo i okruzhaiushchaia sreda*, Moscow, 1973; and his article "Tekhnicheskii progress i zashchita biosfery," *Kommunist* 11(1973).

12. Loren R. Graham, *The Soviet Academy of Sciences and the Communist Party, 1927–1932*, Princeton, 1967, pp. 100–102, 180, 110, 116–233, 127, 131, 132–34, 136–38.

13. "Problema vremeni v osveshchenii Akad. Vernadskogo," *Izvestiia AN SSSR, Otdelenie matematicheskikh i estestvennykh nauk* (hereafter IAN-OMEN), 1932, p. 56.

14. "Po povodu kriticheskikh zamechanii Akad. A. M. Deborina," IAN-OMEN, p. 396.

15. Ibid., p. 404.

16. Ibid., p. 401.

17. Ibid., p. 404.

18. Ibid., p. 401.

19. Ibid., p. 404.

20. "Kriticheskie zamechaniia na kriticheskie zamechaniia Akad. V. I. Vernadskogo," IAN-OMEN, 1933, p. 410.

21. He refers to V. I. Vernadsky, *Ocherki i rechi*, Vol. 2, Petrograd, 1922, p. 21.

22. Deborin, p. 419.

23. Ibid.

24. "O predelakh biosfery," IAN-OMEN, 1937, pp. 9–23.

25. "O metode i soderzhanii vyskazyvanii Akad. V. I. Vernadskogo po filosofii," IAN-OMEN, 1937, p. 25.

26. Ibid., p. 34.

27. Ibid., p. 36.

28. Gumilevskii, pp. 214–27; George Vernadsky, "Bratstvo," p. 233.

29. Mochalov, pp. 169–70; Gumilevskii, pp. 214–27.

30. Vernadsky, "Memoirs," p. 8.

31. Interviews with former associates of Vernadsky who asked not to be cited. Moscow, May and June 1981.

32. Vernadsky to V. P. Volgin, June 1930, AAN, f. 518, op. 3, d. 1952, 1. 9; Vernadsky to V. M. Molotov, 17 February 1932, AAN, f. 518, op. 3, d. 51, 1.2.

33. For example, Vernadsky was elected corresponding member of the French Academy of Sciences in 1928. He had been a member of the Czech and Yugoslav academies of science since 1926 and was an active member of the German Chemical Society, the Geological Society of France, and the mineralogical societies of the United States and Germany; see Stepan G. Korneev, *Sovetskie uchenye—pochetnye chleny inostrannykh nauchnykh uchrezhdenii*, Moscow, 1973. He also frequently presented papers at international congresses and published books and papers in a number of foreign languages, including French, German, English, Czech, and Japanese; see *Otchet o deiatel'nosti AN SSSR za 1929 g.*, Leningrad, 1930, pp. 4–5.

34. *Otchet o deiatel'nosti AN SSSR za 1930 g.*, Leningrad, 1931, p. 6; Mikhail I. Sumgin, "Sovremennoe polozhenie issledovanii vechnoi merzloty v SSSR i zhelatel'naia postanovka etikh issledovanii v blizhaishem budushchem," in *Vechnaia merzlota*, Moscow, 1930; Balandin, *Vernadskii*, pp. 147–50.

35. *Pis'ma V. G. Khlopina k V. I. Vernadskomu*, Moscow, 1961, pp. 1–14.

36. Demian I. Kordeev, "Stanovlenie radiogeologii," *Ocherki po istorii sovetskoi nauki i kul'tury,* Moscow, 1968.

37. V. I. Vernadsky, "O kontsentratsii radiia rastitel'nymi organizmami," *Doklady akademii nauk,* seriia A, Leningrad, 1930; AAN, f. 518, op. 1, d. 137.

38. See David Holloway, "Entering the Nuclear Arms Race: The Soviet Decision to Build the Atomic Bomb, 1939–1945," in *Social Studies of Science,* 11(1981): 159–97.

39. *Otchet o deiatel'nosti AN SSSR za 1934,* Moscow, 1935, p. 112; AAN, f. 518, op. 4, d. 51, 1. 8; op. 2, d. 48, 1. 108; *Otchet o deiatel'nosti AN SSSR za 1935,* Moscow, 1936, p. 168.

40. "Materialy po deiatel'nosti v BIOGEL, 1935–1938," AAN, f. 518, op. 4, d. 51, 11. 33–38; d. 54. See also the article by Dmitrii P. Maliuga in *Pravda,* 13 June 1941, concerning the work of the BIOGEL laboratory.

41. Konstantin P. Florinskii, "100-letie so dnia rozhdeniia akademika V. I. Vernadskogo," *Geokhimiia* 3(1963): 90; AAN, f. 518, op. 4, 1. 554, 11. 33–34; op. 4, d. 80, 1. 1116; Aleksandr P. Vinogradov, "Geokhimicheskii i biogeokhimiia," in *Uspekhi khimii,* 5(1938); and idem, "Geokhimicheskii issledovaniia v raione raspostraneniia urovskoi endemii," *Doklady AN 23* 1(1939).

42. AAN, f. 518, d. 49, 1. 39.

43. V. I. Vernadsky, "O neobkhodimosti vydeleniia i sokhraneniia chistykh tiazhelykh izotopov prirodynykh radioaktivnykh protsessov," *Priroda* 1 (1941): 64.

44. *Trudy BIOGELa,* No. 7, 1937.

45. *Otchet AN za 1929,* pp. 98–99; Mochalov, "Vladimir Ivanovich Vernadsky," p. 516; anonymous, "Memoirs of a BIOGEL worker," manuscript in private collection of a Moscow scientist, shown to me in Moscow, June 1981; *Pravda* 13 June 1941.

46. Mochalov dissertation, "Vladimir Ivanovich Vernadskii," p. 57; I. I. Mochalov, *Vladimir Ivanovich Vernadskii, 1863–1945,* Moscow, 1982, pp. 329–38.

47. *New York Times,* 5 May 1943; AAN, f. 518, op. 2, d. 49, 1. 1.

48. AAN, f. 58, op. 2, d. 49, 1. 1.

49. *Izvestiia* 26 June 1940; *Pravda* 16 July 1940.

50. Letter from Vernadsky to Otto Shmidt, 1 July 1940. AAN, f. 518, op. 3, d. 1817, 1. 11.

51. Vernadsky to Grevs, 2 July 1940, AAN, f. 518, op. 3, d. 562, 1. 1.

52. AAN, f. 518, op. 4, d. 68, 11. 36–38.

53. Ibid., op. 4, d. 68, 11. 13–14.

54. Ibid.

55. Ibid., op. 2, s. 49, 1. 3.

56. Ibid., op. 4, d. 68, 1. 3; op. 2, d. 49, 1. 4; op. 4, d. 68, 11. 6–7.

57. Vernadsky to B. L. Lichkov, 7 September 1940, Vernadsky Museum Collection (VMC), Institute of Geochemistry, Moscow.

58. Holloway, "Entering the Nuclear Arms Race," p. 167.

59. Ibid., p. 170.

60. Ibid., p. 187.

61. Vernadsky to V. L. Komarov, 15 March 1942, AAN, f. 518, op. 2, d. 55, 1. 195; Vernadsky to A. P. Vinogradov, 25 November 1942 and December 1942, op. 2, d. 52, 11. 98, 354; Vinogradov to Vernadsky, 12 December 1942, d. 53, 1. 214; Vernadsky to Komarov, 13 March 1943, op. 2 d. 55, 1. 182. Also see Petr T. Astashenko, *Kurchatov,* Moscow, 1967, p. 130 and Nikolai V. Belov, ed., *D. I. Shcherbakov, zhizn' i deiatel'nost'* Moscow, 1969, pp. 280–81.

62. Vernadsky to Fersman, 27 November 1942, VMC.

63. See Holloway's untitled paper on the Ioffe Institute delivered at the Stanford Conference on Soviet Science and Technology, July 1984.

64. Ibid.

65. Holloway, "Entering the Nuclear Arms Race," p. 175.

66. Igor N. Golovin, *I. V. Kurchatov*, Moscow, 1967, p. 481; Astashenkov, p. 170; see also David Holloway, "Military Technology," in *The Technological Level of Soviet Industry*, ed. Ronald Amann, New Haven, 1977, pp. 451–55. In "Entering the Nuclear Arms Race," Holloway related the story that Stalin summoned Ioffe and Vernadsky to a meeting in the Kremlin in 1942 to discuss the possibility of an atomic bomb, but the story is apparently apocryphal, according to later information Holloway received from Soviet sources. For this apocryphal incident, see pp. 174–75 of the Holloway article. In the summer of 1984, Holloway communicated to me that he had concluded that the Astashenkov book, in which the purported meeting with Stalin is related, is inaccurate on this point. I would agree, having found no evidence in the Vernadsky materials that he ever met personally with Stalin on any topic, either before or during World War II. Vernadsky spent most of the war at a Kazakhstan spa that was maintained by the Academy of Sciences for elderly and infirm members. Beyond that, the Soviet government may not have fully trusted Vernadsky with defense information, given his family ties with the West and his liberal, anti-Marxist background.

67. AAN, f. 518, op. 2, d. 55, 1. 182.

68. Ibid.

69. *D. I. Shcherbakov*, pp. 280–81.

70. Vasilii S. Emel'ianov, "U istokov atomnoi promyshlennosti," *Voprosy istorii* 5(1975): 123–39 and idem, *S. chego nachinalos'* Moscow, 1979.

71. Robert L. Lewis, "Some Aspects of the Research and Development Effort of the Soviet Union," *Soviet Studies* 2 (1972): 153–79.

72. AAN, f. 518, op. 1, d. 325, 11., 1–3; Vernadsky to Fersman, 24 September 1932, 7 March 1933, 1 October 1936, 1 July 1943, VMC; anonymous, "Memoirs of a BIOGEL Worker"; Vernadsky to V. A. Obruchev, 1938, AAN, MO, f. 518, op. 4, d. 62, 11. 3–4; Vernadsky to Komarov, 3 August 1937 and to G. M. Kzhizhanovskii, 10 December 1937, AAN, f. 518, op. 4, d. 51, 11. 23, 21,; op. 3, d. 863, 1. 3.

73. Vernadsky to Fersman, May 1941, VMC.

74. Ibid.

75. Vernadsky to Fersman, 7 March 1933 and 12 July 1937, VMC.

76. Vernadsky to Komarov and Vernadsky to N. G. Sadchikov, head of Glavlit, the censorship bureau, 18 March 1942, AAN, f. 518, op. 2, d. 52, 1. 92.

77. Vernadsky to Fersman, 7 March 1933 and 12 July 1937, VMC.

78. Mochalov dissertation, pp. 528–29; Mochalov, 1982, pp. 304–5. The essay was first published in *Biulleten' Moskovskogo obshchestva ispytatelei prirody, otdelenie geologii* 21(1946): 4–5. The essay has been republished recently in V. I. Vernadsky, *Izbrannye trudy po istorii nauki*, Moscow, 1981, pp. 242–89.

79. Vernadsky to Lichkov, 16 September 1944, AAN, f. 518, op. 3, d. 1756, 1. 30; Mochalov dissertation, p. 514.

80. Vernadsky to Lichkov, 6 September 1944, AAN, f. 518, op. 3, d. 1756, 1. 30.

81. Copies of the complete correspondence between Lichkov and Vernadsky can be found in the VMC.

82. See two manuscripts in BC: M. M. Samygin, "Partiinaia prosloika v AN SSSR," and N. Iefremov "Conditions of Research in Soviet Geology," and also Vernadsky to Lichkov, 7 July 1940, VMC.

83. *Perepiska V. I. Vernadskogo s B. L. Lichkovym, 1918–1939*, Moscow, 1979 and *Perepiska V. I. Vernadskogo s B. L. Lichkovym, 1940–1944*, Moscow, 1980.

84. Anonymous, "Memoirs of a BIOGEL Worker," pp. 5ff; Gumilevskii, *Vernadskii*, Moscow, 1967, pp. 214–27. Another student of Vernadsky's ideas, the biologist Vladimir V. Stanchinskii, developed and began to quantify the concept of energy transfers between trophic levels in an ecosystem a decade before the U.S. biologists G. Evelyn Hutchinson and Ray Lindeman. Stanchinskii, however, came into conflict with the

Michurinists and the followers of Lysenko in the 1930s and was ousted from his laboratory. His work was disrupted and virtually forgotten until the 1970s. See the excellent article by Douglas Weiner, "The Historical Origins of Soviet Environmentalism," in *Environmental Review*, 6(No. 2, 1982): pp. 42–62. Another biologist closely associated with Vernadsky's school in the 1930s, Georgii F. Gauze, also ran into trouble with the followers of Lysenko but managed to survive; see Weiner, "Origins of Soviet Environmentalism," p. 52. Biologists associated with Vernadsky encountered more political problems in this period than geochemists or geologists, perhaps in part because the Lysenko followers had become entrenched in the biological sections of Soviet scientific institutions but had little influence on the physical, chemical, or geological sections. On the efforts of Soviet chemists and physicists to resist Lysenkoism, see Mark B. Adams, "Biology in the Soviet Academy of Sciences, 1953–1965: A Case Study in Soviet Science Policy," in *Soviet Science and Technology: Domestic and Foreign Perspectives*, ed. John R. Thomas and Ursul M. Kruse-Vaucienne, Washington, D.C. 1977, pp. 161–88.

85. AAN, f. 518, op. 1, d. 125, ll. 1–3; see also Vernadsky to N. G. Bruevich, 9 December 1943, AAN, f. 518, op. 1, d. 6, ll. 168–69.

86. Vernadsky to Fersman, 1 July 1943, VMC.

87. AAN, f. 518, op. 1, d. 125, ll. 1–3.

88. Vernadsky to N. N. Luzin, 19 June 1943, quoted in Mochalov, 1970, pp. 169–70.

89. Vernadsky to Komarov, 15 March 1943, AAN, f. 518, op. 2, d. 55, l. 32.

90. See, for example, AAN, f. 518, op. 4, d. 62, l. 3; op. 4, d. 177, ll. 77–78; Vernadsky to V. A. Obruchev, 1938, f. 518, op. 4, d. 62, ll. 3–4.

91. *Khimicheskoe stroenie biosfery zemli i ee kruzheniia*, Moscow, 1965.

92. *Razmyshleniia naturalista*, Moscow, 1977; *Prostranstvo i vremia v nezhivoi i zhivoi prirode*, Moscow, 1975; *Razmyshleniia naturalista*, vol. 2, *Nauchnaia mysl' kak planetnoe iavlenie*, Moscow, 1977.

6. THE LEGACY OF VERNADSKY'S THOUGHT

1. Among his chief scientific heirs in these areas have been A. E. Fersman, A. P. Vinogradov, V. G. Khlopin, Ia. V. Samoilov, N. N. Slavianov, O. Iu. Shmidt, N. G. Kholodnyi, K. A. Nenadkevich, D. I. Shcherbakov, I. E. Lichkov, and their pupils.

2. V. I. Vernadsky, "Mysli o sovremennom znachenii istorii znanii," *Trudy komissii po istorii znanii* 1 (1927).

3. See *Razmyshlenie naturalista*, Vol. I, Moscow, 1975; *Priroda*, 10 (1967): 97–99; Mochalov, pp. 72–170.

4. I. V. Kuznetsov, "Estestvoznanie, filosofiia, i stanovlenie noosfery," *Voprosy filosofii*, 12 (1974): 128–38. Vernadsky had taken a leading part in the struggle to preserve the autonomy and prevent a Communist takeover of the Academy of Sciences after 1927, opposing the election of such Marxist philosophers as A. M. Deborin to the Academy. He defended his stand in a long memorandum in which he asserted that Marxism was an 'outworn philosophy' that held back the development of the sciences. For Vernadsky's role in this dispute and its outcome, see Loren R. Graham, *The Soviet Academy of Sciences and the Communist Party 1927–1932*, Princeton, 1967, pp. 100–102, 108, 110, 116, 123, 127, 131, 132–34, 136–38; George Vernadsky, *Novyi zhurnal* 97, pp. 231–32.

5. Surprisingly, he was even allowed to publish his dissent in 1940, a privilege apparently afforded only to someone of his scientific stature. Even then, the editorial committee of the Academy of Sciences criticized his views in a prefatory note (*Biogeokhimicheskie ocherki 1922–32*, Moscow, 1940, p. 3). In his own preface, Vernadsky wrote, "Recognizing, of course, the deepest significance of philosophy in the life of

humanity and having devoted several years to its study, the author has finally in his own life come to the conclusion that in the given historical moment, philosophy must not have primacy over science in the discussion of scientific questions." (Ibid., p. 8) He went on to say that he thought a new philosophy would eventually emerge, "which inevitably will be a consequence of the scientific movement currently taking place, unprecedented in earlier history. But the philosopher cannot at the present moment show the way to the scientist with usefulness for the latter; he inevitably must at this time occupy second place for a scientific understanding of what is occuring. Philosophy is important and valuable but it does not lead humanity into the scientific realm." (Ibid.)

Vernadsky's preface is dated July 1935, but was not published until 1940. This was the continuation of a polemic between Vernadsky and prominent Marxist philosophers which flared up several times in the publications of the Soviet Academy of Sciences and the Communist Party during the 1930s. See, in particular, D. M. Novogrudskii, "Geokhimiia i vitalizm," *Pod znamenem marksizma*, 7–8 (1931); V. I. Vernadskii, "Problema vremeni v sovremennoi nauke," *Izvestiia Akademiia nauk SSSR, Otdelenii matematicheskikh i estestvennykh nauk*, 4 (1932): 511–41; A. M. Deborin, "Problema vremeni v osveshchenii Akad. Vernadskogo," Ibid., pp. 543–69; V. I. Vernadskii, "Po povodu kriticheskikh zamechanii Akad. A. M. Deborina," Ibid., 3 (1933): 395–419; A. A. Maksimov, "O metode i soderzhanii vyskazyvanii Akad. V. I. Vernadskogo po filosofii," Ibid., 1 (1937): 25–37; V. I. Vernadskii, "O predelakh biofery," Ibid., pp. 1–24.

6. See David Joravsky, *Soviet Marxism and Natural Science 1917–1932*, New York, 1960.

7. Mochalov, pp. 90–97; Kuznetsov, pp. 137–38; *Priroda* 6 (1973), p. 31. I. A. Kozikov, *Filosofskie vozzreniia V. I. Vernadskogo*, Moscow, 1963. Kozikov tries to prove that Vernadsky was, in his methods and thought if not in his language, a dialectical materialist, even though he refused to use such terms to describe his viewpoint. Such attempts to integrate Vernadsky's ideas to the official philosophy go back to the 1920s and ultimately to Lenin's *Materialism and Empirio-criticism* (1909). I think they indicate the seriousness with which Vernadsky's work is taken and his stature in Soviet science. They also indicate the need Communist writers have seen to incorporate such an important scientist within the framework of dialectical materialism. For an article from the 1920s by a Soviet opponent of Vernadsky, arguing against the viewpoint that Vernadsky was an unwitting dialectical materialist, see *Revoliutsiia i kul'tura* 19–20 (1930), p. 160ff.

8. V. I. Vernadskii, *Khimicheskoe stroenie biosfery Zemli i ee okruzhenii*, Moscow, 1965, p. 271.

9. For memoirs to this effect, see especially *Ocherki po istorii geologicheskikh znanii*. Vypusk 11. *Zhizn' i tvorchestvo V. I. Vernadskogo po vospominaniiam sovremennikov*, Moscow, 1963.

10. *Razmyshleniia naturalista*, vol. 2, p. 7.

11. See, in addition to some of the sources cited in note 12 below, *Literaturnaia gazeta*, 24 January 1973, p. 12, and *Vvedenie v geogigienu. Posviashchastesia pamiati akademika V. I. Vernadskogo*, Moscow, 1966. An account in English of Soviet problems with the environment are in Marshall Goldman, *The Spoils of Progress*, Cambridge, England, 1972.

12. For the great increase of articles and books about Vernadsky see *Letopis' zhurnal'nykh statei* before and after 1960. Among his major works published for the first time in recent years, see *Khimicheskoe stroenie biosfery Zemli i ee okruzhenii*, Moscow, 1965; *Priroda* 6 (1973): 30–41; and *Razmyshlenie naturalista*, vol. 1, Moscow, 1974. In addition, large excerpts of his extensive diaries and correspondence have been published. See, for example, *Priroda*, Nos. 10, 11, 12, 1967; *Perepiska A. E. Fersman s V. I. Vernadskim*, Moscow, 1965 and V. G. Khlopin, *Pisma k V. I. Vernadskomu*, Moscow, 1961. For a more complete bibliography of works by and about Vernadsky, see Lev Gumilevskii, *Vernadskii*, Moscow, 1961, 1967; V. I. Vernadsky, *Izbrannye sochineniia*,

vols. I–V, Moscow, 1954–60; I. I. Mochalov, *V. I. Vernadskii—Chelovek i myslitel'*, Moscow, 1970; N. I. Bronskii, *Vernadskii—K. stoletiiu so dnia rozhdeniia*, Rostov, 1963; *Vospominaniia o V. I. Vernadskom, Sbornik*, Moscow, 1963; V. I. Vernadsky, *Biosfera*, Moscow, 1967, 372–74.

13. *Biosfera*, Moscow, 1967 and *Izbrannye sochineniia*, vol. V, 1960; *Izbrannye trudy po biogeokhimii*, Moscow, 1967.

14. *Biosfera*, 22. Most scientists in the West would not consider Vernadsky the sole founder of geochemistry but would divide that role among Vernadsky, F. W. Clarke, V. M. Goldschmidt and their collaborators. For a recent Soviet history of geochemistry in the USSR, see A. P. Vinogradov in *Razvitie nauk o Zemle v SSSR 1917–1967*, Moscow, 1967, p. 70ff.

15. See the correspondence between Dokuchaev and Vernadsky (1888–1900) published in *Nauchnoe nasledstvo. Estestvenno-nauchnaia seriia*, vol. 2, Moscow, 1951.

16. For accounts of Vernadsky's career before 1917, see Alexander Vucinich, *Science in Russian Culture 1861–1917*, Stanford, 1970 and Gumilevskii, op. cit. His chief works before 1914 were *Mineralogiia*, Moscow, 1908, and *Opyt opisatel'noi mineralogii*, St. Petersburg, 1909, 1912. In addition, he published dozens of scientific papers on mineralogy, crystallography, and geochemistry in this period.

17. *Biogeokhimicheskie ocherki*, 1940, p. 6.

18. V. V. Dokuchaev, "Mesto i rol' sovremennogo pochvovedeniia v nauke i zhizni," published in his *Sochineniia*, vol VI, Moscow, 1951, p. 399.

19. "O nauchnom mirovozrenii," *Voprosy filosofii i psikhologii*, No. 65, p. 1409ff.

20. V. I. Vernadsky, "Ob usloviiakh poiavleniia zhizni na zemle," *Izvestiia akademii nauk SSSR*, 1931.

21. R. Balandin, *Vernadskii: zhizn', mysl', bessmertie*. Moscow, 1979, p. 145.

22. See Vernadsky's *O neobkhodimosti issledovaniia radioaktivnykh mineralov Rossiskoi imperii*, St. Petersburg, 1910.

23. *Biogeokhimicheskie ocherki 1922–1930*, Moscow, 1940, p. 96.

24. See Vernadsky's "Izuchenie iavlenii zhizni i novaia fizika," *Izvestiia Akademii nauk SSSR*, 7 seriia, OMEN, No. 3, 1931, pp. 403–37.

25. Vernadsky, "Memoir," pp. 11–12. V. I. Vernadsky's memoirs are in *The Annals of the Ukrainian Academy of Arts and Sciences in the U.S., Inc*.

26. Mochalov, 113–24; V. I. Vernadsky, "Memoir" (see note 25), 12–13; Gumilevskii, 121–62.

27. Fedoseev, 616–18; A. P. Vinogradov, "Vernadskii, V. I.," *The Great Soviet Encyclopedia*, vol. 4, New York, 1974, 611–12.

28. V. I. Vernadsky, "Memoir," 11.

29. Gumilevskii, 124. On the origins and development of Vernadsky's concept of the biosphere, see also *Nauka i zhizn'*, 3 (1974): 40–44 and *Vestnik Leningradskogo universiteta, Seriia biologii*, 9 (1962), vypusk no. 2, 5–21.

30. "Khimicheskii sostav zhivogo veshchestva v sviazi s khimiei zemnoi kory," in *Biogeokhimicheskie ocherki 1922–1932*, Moscow, 1940.

31. Gumilevskii, p. 124.

32. Vernadsky, "Memoir," p. 12.

33. V. I. Vernadsky, "Evoliutsiia vidov i zhivoe veshchestvo," *Priroda*, 3 (1928), 227.

34. Vernadsky, "Memoir," p. 12.

35. Gumilevskii, pp. 125–26.

36. Gumilevskii, p. 126.

37. *New Scientist*, 2 January 1973, p. 15.

38. Gumilevskii, p. 127.

39. "Sur la représentation de la composition chimique de la matière vivante," *C. r. Acad. sci.*, Paris, vol. 179, pp. 1215–17; "La matière vivante et la chimie de la mer," *Revue générale des sciences*, 1924, vol. 35, no. 1, pp. 5–13; no. 2, pp. 46–54; "Sur la

pression de la matière vivante dans la biosphere," *C. r. Acad. sci.*, Paris, 1925, vol. 180, pp. 2079–81; "Sur la portée biologique de quelque manifestations géochimique de la vie," *Revue générale des sciences*, 1925, vol. 36, no. 10, pp. 301–4; "L'autotrophie de l'humanité," *Revue générale des sciences*, 1925, vol. 36 "Sur la multiplication des organismes et son role dans le mécanisme de la biosphere," *Revue générale des sciences*, 1926, vol. 37, no. 23, pp. 661–68; no. 24, pp. 700–708. His published works in Russian on the biosphere include "Zapiska ob izuchenii zhivogo veshchestva s geokhimicheskoi tochki zreniia," *Izvestiia Ross. Akad. nauk*, 6 seriia, 1921, vol. 15, no. 1, pp. 120–23; *Nachalo i vechnost' zhizni*, Petrograd, 1922; *Zhivoe veshchestvo v khimii moria*, Petrograd, 1923; "Ot zhizni v biosfere," *Priroda*, 1925, no. 10–12, pp. 25–38.

40. See the letter from the President of the Carnegie Institution to Vernadsky, 28 April 1923; letter from Vernon Kellog of the National Research Council, 13 December 1923; letter from the Secretary of the British Association for the Advancement of Science, 3 November 1923, all in the Columbia University Archive, Vernadsky Collection, Box 25.4.7.2, Folder 15. A version of Vernadsky's proposal for a biogeochemical laboratory was published in 1923: "A Plea for the Establishment of a Biogeochemical Laboratory," *Transactions of the Marine Biological Station*, 1923, pp. 38–48.

41. Letter of Dana to Vernadsky, 3 April 1923, in Columbia University Archive, Box 25.5.9.1, Folder 15.

42. Letter from Vernadsky to A. E. Fersman, from Paris, 3 March 1923, in *Perepiska A. E. Fersman s V. I. Vernadskim*, Moscow, 1965, p. 421.

43. See note 41.

44. BAR, Columbia University, Vernadsky Collection, Box 25.5.9.1., Folder 12.

45. Ibid.

46. Vernadsky autobiography, BAR, Columbia University.

47. See, for example, letters for 1922–25 in BAR, Columbia University, Box 25.5.9.1 and V. G. Khlopin, *Pisma k V. I. Vernadskomu*, Moscow, 1961.

48. *Biosfera*, Leningrad, 1926 and *La biosphere*, Paris (Alcan), 1929.

49. Vernadsky autobiography, BAR, Columbia University, Vernadsky Collection, Box 31.8.6.1-I-6, Folder I-3-A-a-1-1.

50. Ibid.

51. Quoted in *Ocherki po istorii geologicheskikh znanii*, p. 58.

52. BAR, Columbia University, Box 25.5.9.2.

53. Ibid.

54. Brian Mason, *Principles of Geochemistry*, New York, 1962, p. 233.

55. For an extensive bibliography of these works, see B. L. Lichkov, *Vladimir I. Vernadsky*, Moscow, 1948. Of some 400 works Vernadsky wrote during his lifetime, according to his student, A. E. Fersman, 30 percent dealt with mineralogy, 17 percent with biogeochemistry, 16 percent with other topics in geochemistry, 12 percent with radioactivity, 7 percent with crystallography, 3 percent with soil science, 3 percent with useful mineral deposits, and 11 percent with general questions of science (the history, philosophy, and organization science). A. E. Fersman in *Zapiski Vsesoiznogo mineralogicheskogo obshchestva*, vypusk 1, 1946, p. 75.

56. *Khimicheskoe stroenie*, pp. 5ff.

57. Preface to the French edition, p. ii.

58. Section 103–Section 113.

59. Preface to the Russian edition (1926).

60. Section 8.

61. *Ocherki*, p. 115.

62. See his "Geokhimicheskaia energiia zhizni v biosfere," (1927) in *Biogeokhimicheskie ocherki 1922–32*, Moscow, 1940, p. 127.

63. *Biosfera*, Sections 32–37.

64. *Geochimie in ausgewaehlten Kapiteln*, Leipzig, 1930.

65. *Khimicheskoe stroenie*, p. 294.

66. *Biosfera*, Section 63.

67. See, for example, his "Izotopy i zhivoe veshchestvo," *Doklady AN SSSR*, Series A, December 1926, pp. 245–48 and "O Vlianii zhivikh organismov na izotopicheskie smesi khimicheskikh elementov," *Doklady AN SSSR*, Series A, No. 6, 1931, pp. 141–47.

68. Mason, op. cit., p. 216.

69. *Khimicheskoe stroenie*, p. 294.

70. *Biosfera*, Preface to the Russian edition.

71. A. I. Oparin, *Proiskhozhdenie zhizni*, Moscow, 1924. On the development of Soviet views concerning the origins of life, particularly those of Oparin, see Loren Graham, *Science and Philosophy in the Soviet Union*, New York, 1972, pp. 257–96.

72. *New Scientist*, 6 February 1975, p. 304.

73. *Atmospheric Environment*, vol. 6, 1972, pp. 579–80. See also, P. E. Cloud, *Science*, vol. 160, pp. 729–36; H. D. Holland, "On the chemical evolution of the terrestrial and cytherean atmospheres," in P. J. Brancazio, ed., *The Origin and Evolution of the Atmospheres and Oceans*, New York, 1964, pp. 86–91; and W. W. Rubey in *Geological Society of America Bulletin*, vol. 62, 1951, pp. 1111–47. P. E. Cloud's *Cosmos, Earth and Man* (Yale, 1978), provides a modern synthesis by an American biogeologist on many of the same questions posed by Vernadsky in the 1920s and 1930s.

74. Vernadsky, *Sochineniia*, vol. 5, p. 254.

75. *Biosfera*, section 63.

76. *Khimicheskoe stroenie*, p. 152.

77. Ibid., p. 61.

78. Ibid., p. 234; *Ocherki po istorii geologicheskikh znanii*, p. 127.

79. Bronkii, p. 73; *Khimicheskoe stroenie*, pp. 177–96. In letters to his daughter, Nina Toll 30 June 1938, 15 December 1939 and 30 April 1942, Vernadsky stressed the importance he attached to his work in symmetry and his desire to have that section of his final work on the biosphere translated into English as soon as possible. Letters are in BAR, Columbia University, Box 29.3.5.2., Folder 109.

80. *Khimicheskoe stroenie*, p. 189; "O sostoianiiakh fizicheskogo prostranstva," (manuscript, 1938), published for the first time in *Razmyshlenie naturalista*, vol. I, Moscow, 1975. See also *Khimicheskoe stroenie*, p. 189.

81. See, for example, M. M. Kamshilov, *Evolution of the Biosphere*, Moscow, 1976, and A. E. Presman, *Idei V. I. Vernadskogo v sovremennoi biologii*, Moscow, 1976.

82. V. I. Vernadsky, "Neskol'ko slov o noosfere," *Uspekhi biologii*, vol. 18, Moscow, 1944, pp. 113–120.

83. See E. Le Roy, *L'exigence*, and *Les origines humaines*.

84. *Khimicheskoe stroenie*, p. 247.

85. See, for example, A. P. Vinogradov, "Geochemistry of isotopes," *Vestnik AN SSSR*, no. 5, 1954; *The Elementary Chemical Composition of Marine Organisms*, New Haven, 1953; *Vvedenie v geokhimiiu okeana*, Moscow, 1967 and *Problemy geokhimicheskoi ekologii organizmov*, Moscow, 1974.

86. See, for example, Eugene P. Odum, *Fundamentals of Ecology*, Philadelphia, 1971. A comparison between Odum's basic concepts of the biosphere and biogeochemical cycles and Vernadsky's *Biosphere* and *Geochemistry* will reveal the extent to which modern ecologists are in Vernadsky's debt. As Odum puts it, "The idea of the ecosystem and the realization that mankind is a part of, not apart from, complex 'biogeochemical' cycles with increasing power to modify the cycles are concepts basic to modern ecology . . ." (p. 35). Biogeochemical cycles and man's role in them were important foci of Vernadsky's work between 1917 and 1945.

87. *Khimicheskoe stroenie*, p. 270.

88. Ibid., pp. 323–29.

89. On Vernadsky's views on the search for truth, see I. I. Mochalov, *V. I. Vernadskii*.

90. For example, in his proposal for a biogeochemical lab (BAR, Columbia University, Box 25.4.7.2., Folder V-8, January 1924), he stressed the value of such research in discovering new deposits and new uses for minerals. The economic value of such research also is emphasized in his last work: *Khimicheskoe stroenie,* pp. 271, 272, 274, 298.

91. B. Kozlovskii, ed. *Zapovedniki Sovetskogo Soiuza,* Moscow, 1969, p. 195. My thanks to Douglas Weiner, then a graduate student at Columbia University for pointing this out to me.

92. "Neskol'ko slov o noosfere," in *Biosfera,* p. 357.

93. Mochalov, p. 130.

94. "Zhivoe veshchestvo, 1916–1923," unpublished manuscript in AAN, f. 518, op. 1, ed. khr. 49, 11. 17–18, 19, cited in Mochalov, p. 114. See also *Khimicheskoe stroenie,* p. 351.

95. *Khimicheskoe stroenie,* p. 9.

96. Letter of Hutchinson to the author, 12 July 1978.

97. *Khimicheskoe stroenie,* pp. 204–5.

Bibliography

ARCHIVAL SOURCES

Arkhiv Akademii Nauk, Moskovskaia oblast', Moscow.
Bakhmeteff Archive of Russian History and Culture, Vernadsky Collection, Columbia University, New York.
Vernadskii Kabinet-muzei, Moscow.
Tsentral'nyi, Gosudarstvennyi Arkhiv Oktiabr'skoi Revoliutsii

JOURNALS AND DOCUMENTS

Ekonomicheskaia gazeta
Ekonomicheskii ukazatel'
Front nauki i tekhniki
Genetika
Izvestiia
Izvestiia Akademii Nauk
Narodnoe delo
Novosti dnia
Novyi mir
Novyi zhurnal
Okhrana prirody
Osvobozhdenie
Pravitel'stvennyi vestnik
Pravo
Priroda i sotsialisticheskoe khoziaistvo
Rech'
Revoliutsiia i kul'tura
Russkaia mysl'
Russkie vedomosti
Trudy BIOGELa
Trudy pervogo Vserossiiskogo s"ezda po okhrane prirody
Trudy pervogo Vsesoiznogo s"ezda po okhrane prirody v SSSR
Vestnik Leningradskogo universiteta, Seriia biologii
Vestnik vospitaniia

RUSSIAN AND SOVIET WORKS

Almanakh sovremennykh russkikh gosudarstvennykh deiatelei. St. Petersburg, 1897.
Anonymous. Memoirs of a BIOGEL worker. Private collection, Moscow.
Petr T. Astashenkov. *Kurchatov*. Moscow, 1967.
R. K. Balandin. *Vernadskii: Zhizn', mysl', bessmertie*. Moscow, 1979.
N. A. Balashova. *Rossiiskii liberalizm nachala XX-ogo veka*. Moscow, 1981.
Nikolai V. Belov. *D. I. Shcherbakov, zhizn' i deiatel'nost'*. Moscow, 1969.
A. Blok. *Obzor nauchno-izdatel'skoi deiatel'nosti KEPS, 1915–1920*. Petrograd, 1920.
———. *Kratkii obzor deiatel'nosti postoiannoi kommissii po izucheniiu estestvennykh proizvoditel'nykh sil'*. Petrograd, 1919.

225

N. I. Bronskii, A. Reznikov, et al. *Vernadskii: K stoletiiu so dnia rozhdeniia*. Rostov, 1963.

N. G. Chernyshevskii. *Perepiska Chernyshevskogo s Nekrasovym, Bogoliubovym, i Zelenym, 1855–1862*. Moscow, 1925.

A. M. Deborin. "Problema vremeni v osveshchenii Akad. Vernadskogo," *Izvestiia Akademiia Nauk SSSR, Otdelenie matematicheskhikh i estestvennykh nauk*. 4(1932): 543–69.

V. V. Dokuchaev. *Sochineniia*. Moscow, 1951.

Vasilii S. Emel'ianov. *S chego nachinalos'*. Moscow, 1979.

———. "U istokov atomnoi promyshlennosti," *Voprosy istorii* 5 (1975): 123–39.

P. R. Ferkhmin. *Materialy po otsenke zemel' Nizhegorodskoi gubernii*. Vypusk 8 (1885).

Aleksandr E. Fersman. *Zapiski Vsesoiuznogo mineralogicheskogo obshchestva*, no. 1, 1946.

———. "V. I. Lenin i izuchenie proizvoditel'nykh sil SSSR," in *Problemy mineral'nogo syr'ia*. Moscow, 1975.

E. E. Flint. "Vospominaniia." Manuscript. Kabinet-muzei, Moscow.

Konstantin P. Florenskii. "100-letie so dnia rozhdeniia akademika V. I. Vernadskogo." *Geōkhimiia* 3 (1963).

G. F. Gauze. "Akademik V. I. Vernadskii—osnovopolozhnik sovremennogo ucheniia ob opticheskoi aktivnosti protoplazmy." *Vestnik Akademii nauk SSSR* 2 (1950): 81–86.

Igor N. Golovin. *I. V. Kurchatov*. Moscow, 1969.

V. V. Gorbunov. "Iz istorii bor'by Kommunisticheskoi partii s sektanstvom Proletkul'ta," in *Ocherki po istorii sovetskoi nauki i kul'tury*. Moscow, 1968.

Gosudarstvennyi sovet, Stenograficheskii otchet. Sessiia 1-ia, zasedanie 11. St. Petersburg, 1906.

I. M. Grevs. "V gody iunosti," *Byloe* 12(1918); 16(1927).

Lev Gumilevskii. *Vernadskii*. Moscow, 1961, 1967.

B. S. Itenberg and A. Ia. Cherniak. *Aleksandr Ul'ianov*. Moscow, 1957.

L. V. Ivanova. *Formirovanie sovetskoi nauchnoi intelligentsii, 1917–1927*. Moscow, 1950.

M. M. Kamishilov, *Evolution of the Biosphere*. Moscow, 1976.

P. K. Kazakova. "Vospominaniia." Manuscript. Kabinet-muzei, Moscow.

V. G. Khlopin. *Pis'ma k V. I. Vernadskomu*. Moscow, 1961.

———. "Radii i radioaktivnost' v SSSR." *Nauka i Tekhnika*. Leningrad, 1928.

N. G. Kholodny. "Iz vospominanii o V. I. Vernadskom," in *Pochvovedenie* 7 (1945).

———. *Izbrannye trudy*. 3 vols. Kiev, 1956–57.

———. *Mysli naturalista o prirode i cheloveke*. Kiev, 1947.

G. F. Kir'ianov. *Vasilii Vasil'evich Dokuchaev*. Moscow, 1966.

N. K. Kol'tsov. *K universitetskomu voprosu*. Moscow, 1909.

Stepan G. Korneev. *Sovetskie uchenye—pochetnye chleny inostrannikh nauchnykh uchrezhdeniia*. Moscow, 1973.

A. A. Kornilov. *Sem' mesiatsev sredi golodaiushchikh krest'ian*. Moscow, 1893.

———. "Vospominaniia." Manuscript. Kabinet-Muzei, Moscow.

———. "Vospominaniia o iunosti Fedora F. Oldenburga," *Russkaia mysl'* August 1916, pp. 53–86.

V. G. Korolenko. *Sobranie sochinenii*. Moscow, 1954.

V. M. Korsunskaia and N. M. Verzilin. *V. I. Vernadskii*. Moscow, 1975.

I. A. Kozikov. *Filosofskie vozzreniia V. I. Vernadskogo*. Moscow, 1963.

A. N. Krylov. *Vospominaniia*. Moscow, 1942.

I. V. Kuznetsov, "Estestvoznanie, filosofiia i stanovlenie noosfery," *Voprosy filosofii* 12(1974): 128–38.

A. B. Lapo. *Sledy bylykh biosfer*. Moscow, 1987.

V. L. Lichkov, *Vladimir I. Vernadskii*. Moscow, 1948.

A. A. Maksimov. "O metode i soderzhanii vyskazyvanii Akad. V. I. Vernadskogo po filosofii." *Izvestiia Akademiia Nauk SSSR, Otdelenii matematicheskikh i estestvennykh nauk*. 1(1937): 25–37.

D. A. Miliutin. *Istorii voiny Rossii s Frantsiei v 1799*. St. Petersburg, 1853.

N. M. Mistriakova. "Struktura, nauchnye uchrezhdeniia i kadry AN SSSR, 1917–1940 gg." in *Organizatsiia nauchnoi deiatel'nosti*. Moscow, 1968.

I. I. Mochalov. *V. I. Vernadskii—chelovek i myslitel'*. Moscow, 1970.

———. "Biokosmicheskie vozzreniia V. I. Vernadskogo." *Vestnik akademii nauk SSSR*. 11 (1979): 117–30.

———. *Vladimir Ivanovich Vernadskii, 1863–1945*. Moscow, 1982.

———. "V. I. Vernadskii i L. N. Tol'stoi," manuscript. Vernadsky Collection, Columbia University Archive, Box 31.8.6.1.-I-6.

V. N. Nekhoroshev. "K istorii geologicheskikh uchrezhdenii v SSSR," in *Ocherki po istorii geologicheskikh znanii*. Vol. 7, Moscow, 1958.

Aleksandr V. Nikitenko. *Moia povest' o samom sebe i o tom chemu svidetel' v zhizni byl. Zapiski i dnevnik, 1804–1877*. St. Petersburg, 1904–5.

D. M. Novogrudskii. "Geokhimiia i vitalizm," *Pod znamenem marksizma*. 7–8 (1931).

S. F. Ol'denburg. *Otchet o deiatel'nosti Rossiiskoi Akademii nauk*. Petrograd, 1917.

———. "Vospominaniia akademika S. F. Ol'denburga o vstrechakh s V. I. Leninym v 1887 i 1921 godakh," in *Lenin i Akademiia nauk*.

A. I. Oparin. *Proiskhozhdenie zhizni*. Moscow, 1924.

P. V. Ototskii. "Zhizn' V. V. Dokuchaeva," in *Pochvovedenie*, special issue, 1904.

O. N. Pisarzhevskii. *Fersman*. Moscow, 1959.

———. *Dmitrii Ivanovich Mendeleev 1834–1907*. Moscow, 1959.

V. A. Posse. *Moi zhiznennyi put'. Dorevoliutsionnyi period, 1864–1917 gg*. Moscow, 1929.

A. E. Presman. *Idei V. I. Vernadskogo v sovremennoi biologii*. Moscow, 1976.

N. M. Raskin and I. I. Shafranovskii. "E. S. Fedorov i V. I. Vernadskii (Po materialam arkhiva Akademii nauk SSSR)." *Ocherki po istorii geologicheskikh znanii*, Vol. 8, Moscow, 1959.

A. Rubakin. *N. A. Rubakin: Lotsman knizhnogo moria*. Moscow, 1967.

Sbornik statisticheskikh svedenii po Tambovskoi gubernii. Vol. 15. Tambov, 1890.

S. O. Seropolko. "Vernadskaia." *Zhenskoe delo*. 10 October 1910, pp. 3–6.

I. I. Shafranovskii. *Evgraf Stepanovich Fedorov*. Moscow, 1963.

———. "Raboty V. I. Vernadskogo po kristallografii," *Zapiski vsesoiuznogo mineralogicheskogo obshchestva*. Vypusk l, chast' 75, Moscow, 1946.

A. D. Shakhovskaia. *Kabinet-muzei V. I. Vernadskogo*. Moscow, 1959.

D. I. Shcherbakov. "Institut geokhimii, mineralogii, i kristallogradii im. M. V. Lomonosova." *Nauka i tekhnika*. Leningrad, 1928.

V. V. Shcherbina and V. S. Neapolitanskaia. "Nauchnyi vklad V. I. Vernadskogo v razvitie geologii i geokhimii." *Sovetskaia geologiia* 3 (1963): 1–17.

G. I. Shchetinina. *Universitety i obshchestvennoe dvizhenie*. Moscow.

P. I. Shlemin. "Zemsko-liberal'noe dvizhenie i adresa 1894/95 g.," *Vestnik Moskovskogo universiteta: Istoriia*. 1 (1973): 197–99.

O. M. Shubnikova. "Akademik V. I. Vernadskii i professor Iakov V. Samoilov." *Ocherki po istorii geologicheskikh znanii*. Moscow, 1953.

German Smirnov. *Mendeleev*. Moscow, 1974.

N. Speranskii. *Voznikovnenie Moskovskogo gorodskogo narodnogo universiteta imeni A. L. Shaniavskogo*. Moscow, 1913.

Stenograficheskii otchet, Gosudarstvennaia Duma.

S. P. Timoshenko, *As I Remember*, New York, 1968.

Ol'ga Trubetskaia. *Kniaz' S. N. Trubetskoi: Vospominaniia sestry*. New York, 1953.

V. A. Ul'ianovskaia. *Formirovanie nauchnoi intelligentsii v SSSR, 1917–1937*. Moscow, 1966.

L. V. Vasil'eva. "Vospominaniia." Manuscript. Kabinet-muzei, Moscow.

Nina V. Vernadskaia. *Vospominaniia*. Manuscript. Archives of the Hoover Institution, Stanford, California.

Vladimir I. Vernadskii. *Izbrannye trudy po istorii nauki*. Moscow, 1981.

————. *Biogeokimicheskie ocherki, 1922–1932*. Moscow, 1940.

————. *Biosfera*. Leningrad, 1926; Moscow, 1967.

————. "Blizhaishie zadachi akademicheskoi zhizni," *Pravo*, 19 June 1905.

————. Diary, unpublished. Kabinet-muzei, Moscow.

————. "Evoliutsiia vidov i zhivoe veshchestvo." *Priroda* 3 (1928): 227–50.

————. "The First Year of the Ukrainian Academy of Sciences, 1918–1919." *Annals of the Ukrainian Academy of Arts and Sciences in the US*. 11 (1964–68).

————. "Geokhimiia v Soiuze." *Nauka i tekhnika*. Leningrad, 1928.

————. "Geokhimicheskaia energiia zhizni v biosfere." *Zentralblatt für Mineralogie, Geologie und Paläontologie* 11 (1928): 583–94.

————. "Iz vospominanii." Manuscript. Kabinet-muzei. Moscow, 1943.

————. *Izbrannye sochineniia*. Vols. I–V. Moscow, 1954–60.

————. *Izbrannye trudy po istorii nauki*. Moscow, 1981.

————. "Izotopy i zhivoe veshchestvo," *Doklady AN SSSR*, seriia A. December, 1926: 245–48.

————. "Izuchenie iavlenii zhizni i novaia fizika," *Izvestiia Akademii Nauk SSSR*, 7 seriia, OMEN, No. 3, 1931, pp. 403–37.

————. *Khimicheskii sostav zhivogo veshchestva v sviazi c khimiei zemnoi kory*. Petrograd, 1922.

————. *Khimicheskoe stroenie biosfery Zemli i ee okruzhenii*. Moscow, 1965.

————. "Khod zhizni v biosphere," *Priroda*, 1925, nos. 10–12: 25–38.

————. *La biosphere*. Paris, 1929.

————. "La matière vivante et la chimie de la mer," *Revue générale des sciences,* 35 (no. 1, 1924): 5–13; 35 (no. 2, 1924): 46–54.

————. "L'autotrophie de l'humanité," *Revue générale des sciences* 36 (1925).

————. *Mineralogiia*. Moscow, 1908.

————. *Nachalo i vechnost' zhizni*. Petrograd, 1922.

————. "Neskol'ko slov o noosfere," *Uspekhi biologii*, Vol 18, Moscow, 1944: 113–20.

————. "O fosforitakh Smolenskoi gubernii," *Trudy Vol'nogo ekonomicheskogo obshchestva*. 11 (1888): 263–94.

————. *O gruppe sillimanita i roli glinozema v silikatakh*. Moscow, 1891.

————. "O nauchnom mirovozzrenii." *Voprosy filosofii i psikologii*. 65 (1902).

————. *O neobkhodimosti issledovaniia radioaktivnykh mineralov Rossiskoi imperii*. St. Petersburg, 1910.

————. "O polimorfizme kak obshchem svoistv materii," in *Uchenye zapiski Mosovskogo universiteta*. Otdelenie estestvenno-istoricheskii, vypusk 9, 1892.

————. "O predelakh biosfery," *Izvestiia Akademii Nauk. otdelenii matematicheskikh i estestvennykh nauk*. 1 (1937): 1–24.

————. "O vlianii zhivykh organismov na izotopicheskie smesi khimicheskikh elementov," *Doklady AN SSSR*, seriia A. 6(1931): 141–47.

————. "O zemnykh aliumofosfornykh i aliumosernykh analogakh kaolinovykh aliumosilikatov," *Doklady Akademii nauk*. 18 (nos. 4–5, 1938): 621ff.

————. *Ob osnovaniiakh universitetskoi reformy*. Moscow, 1901.

————. "Ob usloviiakh poiavleniia zhizni na zemle." *Izvestiia akademii nauk SSSR*. 1931: 633–53.

————. *Ocherki i rechi Akad. V. I. Vernadskogo*. 2 vols. Petrograd, 1922.

————. *Opyt opisatel'noi mineralogii*. St. Petersburg, 1909, 1912.

———. *Osnovy kristallografii*. Moscow, 1903.

———. *Perepiska V. I. Vernadskogo s B. L. Lichkovym, 1918–1939*. Moscow, 1979.

———. *Perepiska V. I. Vernadskogo s B. L. Lichkovymm 1940–1944*. Moscow, 1980.

———. *Pis'ma V. I. Vernadskogo A. E. Fersmanu*. Moscow, 1985.

———. "Po povodu kriticheskhikh zamechanii Akad. A. M. Debornina." *Izvestiia Akademiia Nauk SSSR, Otedelenii matematicheskikh i estestvennykh nauk*. 3(1933): 395–419.

———. "Prilozhenie," *Osvobozhdenie*. 78/79 (1905): 1–14.

———. "Problema vremeni v sovremennoi nauke," *Izvestiia Akademiia Nauk SSSR, Otdelenie matematicheskikh i estestvennykh nauk* 4(1932): 511–41.

———. *Razmyshlenie naturalista*. 2 vols. Moscow, 1975.

———. "Sur la multiplication des organismes et son rôle dans le mécanisme de la biosphere," *Revue générale des sciences* 36 (no. 23, 1926): 661–68; 36 (no. 24, 1926): 700–708.

———. "Sur la portée dialogique de quelque manifestations géochimique de la vie," *Revue générale des sciences* 36 (No. 10, 1925): 301–4.

———. "Tri resheniia," *Poliarnaia zvezda* 14 (1906).

———. "Tri zabastovki," *Russkie vedomosti*, 9 August 1905.

———. "Zadacha dnia v oblasti radiia," *Izvestiia Akademii nauk*, 5 (No. 1, 1911): 61–72.

———. "Zapiska o neobkhodimosti organizatsii khimicheskogo izucheniia organismov," *Protokoly zasida' fiz.-mat. viddilu Ukr. AN u Kiivi*, 1918. Vypusk 1, Kiiv, Ukr. AN, 1919, pp. 43–45.

———. "Zapiska ob izuchenii zhivogo veshchestva s geokhimicheskoi tochki zreniia," *Izvestiia Ross. Akademiia Nauk*, 6 seriia. 15 (no. 1, 1921): 120–23.

———. *Zapiska ob otnoshenii Moskovskogo universiteta k "Moskovskim vedomostiam" i k universitetskoi tipografii*. Moscow, 1905.

———. *Zhivoe veshchestvo v khimii moria*. Petrograd, 1923.

A. P. Vinogradov. *Razvitii nauk v Zemle v SSSR 1917–1967*. Moscow, 1967.

———. *Chelovek, obshchestvo i okruzhaiushchaia sreda*. Moscow, 1973.

———. "Vernadskii, V. I.," *The Great Soviet Encyclopedia*, Vol. 4, New York, 1974, pp. 611–12.

———. "Geochemistry of isotopes." *Vestnik AN SSSR*, 5(1954).

———. "Geokhimiia i biogeokhimiia." *Uspekhi khimii* 5(19-8).

———. "Geokhimicheskii issledovaniia v raione rasprostraneniia urovskoi endemii." *Doklady AN 23* 1(1939).

———. "Tekhnicheskii progress i zashchita biosfery." *Kommunist* 11 (1970).

Zhizn' i tvorchestvo V. I. Vernadskogo po vospominaniiam sovremennikov, in *Ocherki po istorii geologicheskikh znanii* 11 (1963).

WESTERN WORKS

Mark B. Adams. "Biology in the Soviet Academy of Sciences, 1953–1965: A Case Study in Soviet Science Policy." *Soviet Science and Technology: Domestic and Foreign Perspectives*. John R. Thomas and Ursula M. Kruse-Vaucienne, eds. Washington, D.C., 1977.

Garland Allen. *Life Science in the Twentieth Century*. New York, 1975.

Ronald Amann, ed. *The Technological Level of Soviet Industry*. New Haven, 1977.

Kendall E. Bailes. *Technology and Society Under Lenin and Stalin*. Princeton, 1978.

Alain Besançon. *Education et société en Russie dans le deuxième tiers du XIXe siècle*. Paris, 1974.

Daniel Brower. *Training the Nihilists: Education and Radicalism in Tsarist Russia*. Ithaca, 1975.

Stephen G. Brush, "Scientific Revolutionaries of 1905: Einstein, Rutherford, Chamberlin, Wilson, Stevens, Binet, Freud," in Mario Bunge and William R. Shea, eds., *Rutherford and Physics at the Turn of the Century*. New York, 1985.

P. E. Cloud. *Cosmos, Earth and Man*. New Haven, 1978.

I. Bernard Cohen. *Revolution in Science*. Cambridge, Mass., 1985.

Dictionary of Scientific Biography. New York, 1972.

Theodosius Dobzhansky. "Memoirs." Oral History Collection, Columbia University Library, Vol. I.

Herbert Dingle. "A Hundred Years of Spectroscopy," *The British Journal for the History of Science*. 1(June 1962).

Patrick P. Dunn. "Childhood in Imperial Russia," in Lloyd de Mause, ed., *The History of Childhood*. New York, 1974.

Terence Emmons. *The Russian Landed Gentry and the Peasant Emancipation of 1861*. Cambridge, Mass., 1968.

———. *The Formation of Political Parties and the First National Elections in Russia*. Cambridge, Mass., 1983.

Sheila Fitzpatrick. *The Commissariat of Enlightenment: Soviet Organization of Education and the Arts under Lunacharsky, October 1917–1921*. London, 1970.

Elizabeth Garber. "Molecular Science in Late Nineteenth Century Britain," *Historical Studies in the Physical Sciences*. 9 (1978):283–84.

Loren Graham. *Science and Philosophy in the Soviet Union*. New York, 1972.

———. *The Soviet Academy of Sciences and the Communist Party 1927–1932*. Princeton, 1967.

Sidney Harcave. *First Blood: The Russian Revolution of 1905*. London, 1965. New York, 1964.

Samuel P. Hays. *Conservation and the Gospel of Efficiency*. Cambridge, Mass., 1959.

David Holloway. "Entering the Nuclear Arms Race: The Soviet Decision to Build the Atomic Bomb, 1939–1945," in *Social Studies of Science*. 11 (1981): 159–97.

David Joravsky, *Soviet Marxism and Natural Science 1917–1932*. New York, 1960.

———. *The Lysenko Affair*. Cambridge, Mass., 1970.

Samuel Kassow, "The Russian University in Crisis, 1899–1911," doctoral dissertation. Princeton University, 1976.

Edward J. Kormondy. *Concepts of Ecology*. Englewood Cliffs, N.J., 1969.

Henri Louis Le Chatelier. *Kremnezem i silikaty*. Leningrad, 1929.

E. Le Roy. *L'exigence idéaliste et la faite de l'évolution*. Paris, 1927.

———. *Les origines humaines et l'évolution d'intelligence*. Paris, 1928.

George Leggett. *The Cheka: Lenin's Political Police*. Oxford, 1981.

Robert L. Lewis. "Some Aspects of the Research and Development Effort in the Soviet Union." *Soviet Studies* 2 (1972): 153–79.

James McClelland. *Autocrats and Academics: Education, Culture, and Society in Tsarist Russia*. Chicago, 1979.

William McGucken. *Nineteenth-Century Spectroscopy: Development of the Understanding of Spectra, 1802–1897*. Baltimore, 1969.

Roberta Manning. *The Crisis of the Old Order in Russia*. Princeton, 1983.

Brian Mason. *Principles of Geochemistry*. New York, 1962.

Zhores Medvedev, *The Rise and Fall of T. D. Lysenko*. New York, 1969.

Paul Miliukov. *Political Memoris, 1905–1917*. Ann Arbor, 1967.

Eugene P. Odum. *Fundamentals of Ecology*. Philadelphia, 1969.

Walter M. Pintner and Don K. Rowney, eds. *Russian Officialdom: The Bureaucratization of Russian Society from the Seventeenth to the Twentieth Century*. Chapel Hill, 1980.

Richard Pipes. *Struve: Liberal on the Left, 1870–1905*. Cambridge, Mass., 1970.

Marc Raeff. *Origins of the Russian Intelligentsia: The Eighteenth Century Nobility.* New York, 1966.

Richard G. Robbins, Jr. *Famine in Russia 1891–1892: The Imperial Government Responds to a Crisis.* New York, 1975.

Ernst Scheibold. *Osnovnye idei geokhimii.* Leningrad, 1933.

Alfred E. Senn and Nicholas Rubakin. *A Life for Books.* Maine, 1977.

Allen Sinel. *The Classroom and the Chancellery: State Educational Reform in Russia under Count Dmitry Tolstoy.* Cambridge, Mass., 1973.

Rene Taton, ed. *Science in the Twentieth Century.* New York, 1964.

Gerard Turner. *Nineteenth Century Scientific Instruments.* Berkeley, 1983.

George Vernadsky. "O rode Vernadskikh," unpublished manuscript dated 29 July 1936. Bakhmeteff Archive, Columbia University.

———. "Bratstvo 'Priiutino'," *Novyi zhurnal,* 93 (1968): 148–70.

———. "The Prijutino Brotherhood." Bakhmeteff Archive, Columbia University.

Alexander Vucinich. *Science in Russian Culture.* 2 vols. Stanford, 1970.

Douglas Weiner. "The Historical Origins of Soviet Environmentalism." *Environmental Review* 6 (1982): 42–62.

Index

29.50